Grassroots Stewardship

Grassroots Stewardship

Sustainability Within Our Reach

F Stuart Chapin, III

OXFORD
UNIVERSITY PRESS

Oxford University Press is a department of the University of Oxford. It furthers
the University's objective of excellence in research, scholarship, and education
by publishing worldwide. Oxford is a registered trade mark of Oxford University
Press in the UK and certain other countries.

Published in the United States of America by Oxford University Press
198 Madison Avenue, New York, NY 10016, United States of America.

CIP data is on file at the Library of Congress
ISBN 978–0–19–008119–5 (hbk.)

1 3 5 7 9 8 6 4 2

Printed by Sheridan Books, Inc., United States of America

Contents

SECTION 2 GRASSROOTS ACTIONS

Prologue: My Journey to This Book

As kids, my friends and I often walked along a small stream in the woods. We teased fish with sticks, watched turtles shrink inside their shells, picked wild grapes within our reach, and sucked the sweetness from yellow honeysuckle blossoms. One day, the trees on the neighboring lot were cleared, and carpenters built a skeleton of house walls that became our temporary jungle gym. On a more distant hillside they cut down the entire forest. Earthmovers as big as houses, or so they seemed, erased the streams, reshaped the land, and created roads bordered by neatly arrayed house lots of red clay soil. These changes were fun for a little boy to watch, but I didn't recognize their incompatibility with my pleasures in the woods.

During trips to my grandmother's house in the Pacific Northwest, my family often stopped to watch Yakima Indians net salmon from rickety platforms above Celilo Falls on the Columbia River (Figure P.1). When we stopped there in 1957, the falls were gone. The floodgates of the newly constructed Dalles Dam had closed the previous March. Within hours, Celilo Falls and its thousands of years of Native American fishing were erased forever—another casualty of "human progress."[1]

During high school, my church youth group often discussed social, ethical, and religious dimensions of life around us in central North Carolina— including the racial segregation of local schools and neighborhoods. Charlie Jones, our locally grounded, globally thoughtful pastor, spoke of the barriers that community leaders faced in trying to bring greater equity to segregated life in Chapel Hill. As my sister and I brought these discussions home, we discovered that my mother's participation in the League of Women Voters and my father's teaching of regional planning at the university were not just "things that grown-ups do," as I had always assumed. Instead, these activities grew out of social conscience. I began to realize that people weren't obliged to accept everything around them as an inevitable reality. Even my own family was trying to change things in ways that had been invisible to me.

I began to participate timidly in marches and picketing of segregated businesses and became friends with African American kids I had never met. As we hung out on the steps of the black Baptist church, I came to understand segregation from their perspective. Several of us were arrested for protesting

Figure P.1 Yakima Indians netting salmon at Celilo Falls (top) before its elimination by the Dalles Dam (bottom).

the Chapel Hill Merchants' Association support for segregated businesses (Chapter 9). The police hauled us out of the building and took us to jail. Although my parents were upset that I was arrested, they understood why I was there.

Marches, sit-ins, and town-hall meetings raised the visibility of segregation, so Chapel Hill leaders felt obliged to address it. Today, Chapel Hill has a very different social and racial environment from when I grew up. Although undercurrents of discrimination persist, I learned that taking political action could make a difference.

After college, my wife Mimi and I joined the Peace Corps to train teachers in Colombia. Our teaching methods required students to think, not just memorize. In an ideal world, a focus on ideas would invigorate informed citizenship and foster democracy. Unsurprisingly, we didn't achieve these lofty goals. And we were often the ones who learned. Some of our partner teachers had a knack for teaching concepts, despite a rigid curriculum and a poorly funded school system. I was impressed by the ability of engaged people, working within their system, to be effective agents of change.

Now for the science side of my life. As a high-school student, I had no idea what I wanted to do when I grew up. My first biology course, taken as a college freshman, was taught by an inspiring group of professors. My favorite was Bill Denison—the green giant whose unpretentious green work clothes and infectious enthusiasm for plants reminded me of a kid who never grew up. Every morning, his laugh echoed through the hallways of the biology building as he talked with students. On field trips, he showed us how biology played out in the forests around campus. He taught me the names and habits of trees based on winter buds that revealed new secrets as the buds burst into spring. We learned the shapes and structures of flowers that hinted at each plant family's evolutionary history. I joined Dr. Denison and two other students on a trip to Costa Rica to search for new species of cup fungi. We explored their hidden habitats near waterfalls and in mountain peat bogs and rainforests. I learned that plants held keys to many intricate puzzles that were both exciting and fun to explore. These intellectual mysteries in enjoyable places launched me toward a career in ecology, but I felt no connection between this path and my social concerns.

Throughout my professional career, I repeatedly found that the topics I studied would be more interesting if placed in a broader context. During my PhD research, I studied temperature effects on nutrient absorption by plants that grew along a thermal gradient between Barrow, Alaska, and Death Valley, California—two places near the temperature limits of life found in the United States. I learned that, along this gradient, plants had adjusted, both physiologically and genetically, to their thermal environment. Plants had the capacity to grow where they did! Not too surprising, in retrospect. But what did surprise me was that the physiological capacity of roots to absorb nutrients didn't really matter because plants scavenged almost all the nutrients they encountered. Plants were limited by the rate at which soil microorganisms released nutrients into the rooting zone. I realized that I would never understand the controls over plant growth unless I also studied plant interactions with soil microorganisms. So, I became an ecosystem ecologist. I study environmental effects on the cycling of nutrients, carbon, and water among soils, plants, and animals.

But this framework was also too restrictive. Climate and patterns of disturbance are changing so quickly that many species now encounter conditions never seen during their evolutionary history. Human impacts on nature are shaping the future of life on Earth. Only by studying the interactions between people and the rest of nature—as an integrated system—can I hope to understand how and why the world is changing. So, that's what I do. An understanding of the linkages between people and nature provides society with the information needed to shape these interactions for the benefit of both. It also provides me with a starting point for taking personal actions that, in my small way, can shape the future of our planet. At long last, my joy in studying nature has merged with my satisfaction in addressing issues that matter to society.

In September 2016, Adele, my 9-year-old granddaughter, who lives in Wales, brought this global challenge into personal focus. The Welsh government had asked primary schools to select child ambassadors to advise it on the rights of children. Adele's class assignment was to list the rights that children thought were important. The United Nations' list of children's rights,[2] which emphasizes protection of children from the evils of the world, didn't resonate with Adele. Instead her "Manifesto on the Rights of Children" began with "the right to good health, which needs healthy food, good water, a clean environment, and special treats once in a while." She added "the right to explore, learn new things, have fun, and be the best that we can, even if this means taking risks—like climbing trees and meeting new people."

If society continues to degrade Earth's environment, Adele's generation will be denied the environmental and societal rights that she identified. Equally important, the spirit of possibilities and empowerment that Adele captured in her manifesto is missing from most public discourse.

This book is written for people who want to help transform a planet in peril to one where society and nature can flourish. This is not a book about the seriousness of problems or about what *should* be done. Instead, it is a book about what *has been* and *can be* done and a strategy for tangible progress. My goal is to shift society's discourse about our planet from one of deep despair about problems to action for solutions.

On the one hand, society has made astounding progress: democracy, health, and an explosion of technology. On the other hand, society has also left in its wake ecosystem degradation, climate warming, and society's isolation from the nature that sustains it. We are in the midst of an unprecedented global experiment, with the fate of humanity and nature hanging in the balance.

Classical economic theory assumes that each person is a pawn that seeks to fulfill its own needs and desires as it is moved by the invisible hand of market forces on a global chessboard. I'm more optimistic than that. As a scientist, I've studied people's roles in both the ecological and social dimensions of transformational change. Many of these changes depend crucially on the actions of individuals—both ordinary and extraordinary. I'm confident that each person *can* make a difference if he or she thinks carefully and acts with intention. In the process, nature and society can flourish together.

This brings me to the start of my story.

SECTION 1

INTEGRATING THE NEEDS OF NATURE AND SOCIETY

1
Shaping the Future Wisely

We have only one Earth. How can we shape its future wisely?

People's Links with Nature

Only three times in Earth's history have biological events deeply altered Earth's environment and chemistry. This first happened 2–3 billion years ago when tiny aquatic organisms—blue-green algae—discovered how to split molecules of water and carbon dioxide to produce organic compounds and oxygen.[1,a] This "Great Oxygenation Event" changed forever the metabolism of the planet. Organisms could then capture more energy through the reaction of oxygen with organic compounds to fuel life's processes compared to Earth's earlier oxygen-free environment. Atmospheric oxygen also generated ozone. This compound shielded Earth from ultraviolet radiation that had previously modified the chemistry of life's genetic code (DNA) in unpredictable ways. These radical environmental changes altered the structure and dynamics of Earth's ecosystems and eventually caused a decline in those ecosystems (**stromatolites**)[b] that housed the blue-green algae responsible for this event. This triggered Earth's first mass extinction.[2]

The second Earth-changing event amplified the Great Oxygenation Event. This occurred 400–450 million years ago when plants invaded land.[3] These plants became so abundant that they increased oxygen concentration to near-modern levels, providing enough oxygen and plant material to support both abundant terrestrial animal life and wildfire.

The third Earth-shaking biological event was human burning of fossil fuels. Scientists argue about when human impacts were first detectable at the planetary scale. Perhaps the human imprint went global when agriculture began

---

[a] Numbered notes and references in the back of the book provide evidence for statements in the text and are shown as superscript numbers.

[b] Technical and other important words in this book are shown in **bold** and are defined in the Glossary in the back of the book.

5,000 years ago or perhaps shortly after the Industrial Revolution began in about 1750, as people began using coal, oil, and gas to support industrial production. However, only since about 1950 did the burning of fossil fuels ramp up enough to change Earth's metabolism in ways that demonstrably altered Earth processes.[4] This short time interval is less than the blink of an eye in geologic terms. It's much too short a time to embed humanity's addiction to fossil fuels in the genetic code of our species. Only since 1990 has the scientific community fully recognized the magnitude and potentially serious consequences of human impacts on our planet.

It's not too late to fix this problem, but society needs to act quickly (Chapters 4 and 10). Understanding and realigning the interactions between people and the rest of nature is the greatest and most urgent societal challenge of our time.[5] This global issue came home to me most compellingly in three unexpected places: Huslia, Alaska; Madison, Wisconsin; and Stockholm, Sweden.

The airport thermometer read −20°F on a January morning in 2004 as I walked to the small plane that would fly me from Fairbanks to Huslia, an Athabascan Indian village of about 300 people in northwest Alaska. A fringe of pastel orange marked the horizon, bordered by angled mountains and spruce-frosted hills. Sunrise blended imperceptibly with sunset, scarcely interrupted by the short arctic day, as the plane droned over pink snow and purple forests.

The community of Huslia had invited five scientists to a workshop to discuss climate changes and impacts. Huslia was even colder than Fairbanks that day—somewhat incongruous for a workshop on climate warming. Cold snow crunched and squeaked underfoot as the five of us walked from our temporary quarters in the school to the community hall, the site for our workshop. This log building was as cold inside as out. Orville Huntington, the local wildlife biologist, had just lit a fire in the barrel stove. While the building warmed, Orville walked with us to the edge of the village to meet his aunt and uncle, Catherine and Steven Attla—two respected community elders.

Catherine welcomed Orville and invited us in for tea. The house was warm, and every surface was covered with half-finished projects—Catherine's beadwork on the table and, in a corner, the beaver that Steven had trapped that morning. After some general chit-chat, Orville told his aunt Catherine that we had come to Huslia for the climate-change workshop. She commented how much the climate had warmed since her childhood and the many ways in which this warming affected their lives, a precursor of comments our visiting group would hear repeatedly during the workshop.

Catherine became more animated as she shifted the conversation to the beaver that Steven had trapped. She pointed out that this was the last beaver he would take from its lodge this winter to ensure that beaver would still be there next year. "We use every part of the beaver. We don't waste anything. We have to show respect for the beaver, or we won't have luck." Respect for nature—including society's dependence on nature—is an Athabascan moral code that would become a central theme of our workshop.

After a day and a half of workshop conversations about climate-change observations, a wizened Athabascan elder rose and said, "It all started when they put a man on the moon." He went on to describe how the weather turned crazy after that. My scientist brain assumed the elder must have mistaken the correlation between the moon landing and climate warming for a causal relationship. But was that his message? His comment seemed to stimulate rather than derail the conversation about climate change. That evening, back at the school, our visiting group talked late into the night about how the elder's comment related to the day's climate-change discussions.[6]

As I later learned, Athabascans traditionally convey their most important points not directly but through stories, riddles, and allegories that the listener is expected to untangle. On many occasions, my scientific questions have been answered with stories. For people who have relied on oral history for thousands of years, stories are more compelling than sets of facts.

Perhaps the elder intended to convey the importance of showing respect for nature—especially for sacred objects like the moon, implying that modern life has run amok because of Western society's rash use of technology to defile nature. The belief that people must respect nature and live in balance with it is a theme I've heard repeatedly in Alaskan villages—not only because such behavior is good for nature but also because human survival depends on it. Not everybody in Alaskan villages abides by this deep moral code framed in Athabascan and other Alaska Native traditions. However, some do, which provides a moral compass that points to a path forward.

Catherine grew up before there were any schools in rural Alaska. Her education came from life on the land and stories told around the fire by her grandfather, Huslia's last medicine man. She taught her children and grandchildren beadwork skills and wrote down many of the stories she learned as a child. But she realized these steps were not enough. At the workshop, Catherine described how recent warming caused bears to emerge early from hibernation, before food was available. "I don't want to talk about bears because women in our culture are never supposed to even say that word. However, someone needs to tell this story. It's all connected." Others at the workshop described how winter was no longer cold enough to thicken the river ice for safe winter travel. Everyone in the village had lost friends or family when snowmobiles went through thin ice. At the end of the workshop, we walked silently back to the landing strip to return to Fairbanks. Each person's frozen words and thoughts were silhouetted against the orange midday sunrise, reminding me of Inuit leader Sheila Watt-Cloutier's declaration of northern peoples' right to be cold.[7]

Like Catherine Attla, Aldo Leopold spent most of his life on the land. In winter 1935, after moving from the western United States to Madison, Wisconsin, he walked down a sand-blown track, looking for land for his family's hunting camp. He bought a worn-out farm and rebuilt its dilapidated chicken coop ("The Shack") to serve as a base for family weekends and for his long walks through abandoned farmlands and second-growth forests. Here, he reflected on the many ways in which modern society had fundamentally disconnected itself from nature by degrading the land. In a collection of essays entitled *A Sand County Almanac*, published in 1949, Leopold articulated a land ethic based on respect and care for the land that became a foundation of modern conservation in the United States and beyond.[8]

Inspired by Leopold's land ethic, a group of private landowners and investors formed the Sand County Foundation in 1967 to support restoration on private lands.[9] Their goal was to discover approaches and motivate individuals to provide leadership for conservation on private lands.

In 2010, as a member of the foundation's advisory committee, I learned first-hand about their philosophy and innovative restoration approaches. The foundation's research showed that farmers could greatly reduce nutrient pollution by placing woodchips coated with nutrient-absorbing microbes in the path of groundwater that drained farm fields. In this way, nutrients supported the growth of decomposer microbes that kept nutrients on the farms rather than flowing downriver to the Mississippi delta, where excess nutrients create a fish-killing dead zone. If nutrient runoff were eliminated from farms along small streams, which make up only a tiny proportion of the Mississippi basin, nutrient input to the Mississippi delta could be reduced by 20%–50%.[10]

The Sand County Foundation also gives awards to farmers and ranchers who have found innovative ways to improve their lands, while turning a profit. These people often become role models for their neighbors.[11] Since private lands account for 75% of the land area of the contiguous United States, effective management on private lands can have large-scale impacts. For me, the biggest surprise was that a conservative political philosophy of empowering landowners to solve their own problems without involving government was so effective at spreading a conservation ethic across private lands.

I was also curious to know whether city people, whose lives are encased in infrastructure, were aware of their connections with the rest of nature. In 2007, I visited the Stockholm Resilience Center (SRC) and the Beijer Institute of Ecological Economics, where I worked with colleagues on a book about ecosystem stewardship.[12] As a break from writing, I often walked among nearby allotment gardens. These gardens had been established in the early 1900s to improve conditions for working-class people who had moved to the city for work.[13] The gardens are now tended by 23,000 Stockholm residents—2.4% of Stockholm's population.

The buzz of bees permeated the brightly flowered gardens, occasionally punctuated by laughs and conversations among adjacent gardeners. The bees support production of flowers, fruits, and vegetables—as well as feeding small birds—in both individual allotments and other urban lands across the city. Garden gossip connected allotment gardeners in a social network grounded in respect for nature and memories of rural roots. Each garden's small shed was a miniature home away from home, with all the basics for sunset picnics. During World Wars I and II, allotment gardens supplied European cities with vegetables at a time when food was scarce—a stark reminder of people's dependence on nature.

Through coffee-room conversations at the SRC and Beijer, I learned that Stockholm's allotment gardens are part of a bigger picture. Stephan Barthel, Johan Colding, Carl Folke, and others have studied these gardens as part of the history and dynamics of human-nature linkages in the Stockholm region.[14]

People have lived in what is now Stockholm's National Urban Park since the glaciers disappeared 3,500 years ago. During the Viking era, people farmed the land. In 1540, the Swedish king took control of the land and built fences to exclude farmers. His grasslands supported livestock and deer, while forests provided oak to build the royal navy. In about 1700, the fences were removed; and, over time, the park became Stockholm's main recreational area. Industrialization in the late 1800s led to the allotment gardens. By 1970, urban sprawl covered a third of the park with pavement and buildings. In the 1980s, plans were released for more extensive development. The resulting public outcry led to establishment of the National Urban Park in 1995.

Birdwatching groups, sports clubs, and allotment garden associations are among the 69 groups that use the park today.[15] Fifty of these groups, involving 18% of Stockholm's residents, played a key role in securing protection for the park and engaging in its management. Both the diverse land-use history of the park and its protection for nearly 500 years largely account for its unusually high species diversity: 75% of the region's species are found within this park that occupies only 1% of the region's land area.

Stockholm's National Urban Park is connected to rural nature through a spiderweb of green spaces.[16] Other green spaces that are integral to the city's social and ecological fabric include tree-lined streets, small parks and playgrounds, gardens and lawns, cultural parks, and cemeteries. Birds, seeds, pollinators, and people link these patches of urban nature. Stockholm residents clearly value and regularly interact with the nature on their doorsteps.

But how can people's local concerns spread beyond their own community? In 2004, educator Rob Hopkins asked his students to develop a plan for their home town of Kinsale, Ireland. Their goal was a plan that would allow the town to thrive in a future when oil might become prohibitively expensive.[17] The Kinsale town council adopted the student plan as a framework to work toward energy independence. This transition-town concept implemented by the Kinsale Council emphasizes use and reuse of renewable resources and less dependence on fossil fuels.

I'm absolutely amazed that, within 10 years, 1,130 transition initiatives had sprung up in 43 countries to support sustainable planning by neighborhoods, communities, and cities![18] They emphasize general resilience (maintaining a diversity of options) to allow nimble responses as an uncertain future unfolds. Similarly, Local Governments for Sustainability, a global network of 1,700 cities, towns, and regions that serve 25% of the global urban population, has committed to a sustainable future by developing green, resource-efficient economies.[19] These initiatives show that many communities and cities recognize that the needs of people and the rest of nature are mutually supportive. Global action for sustainability is not out of reach.

Many powerful organizations and sets of rules that guide people's behavior (**institutions**), such as governments, corporations, and churches, now seek to reduce human impacts on nature through global initiatives. The World Business Council for Sustainable Development (WBCSD), for example, represents 200 transnational corporations that account for 10% of the global gross domestic product.[20] WBCSD designs strategies that enable its member corporations to sustainably harvest raw materials and still be profitable (Chapter 8). Pope Francis of the Catholic Church has reached out to every person on Earth to act as a steward of God's Creation, to care for our world, and to not "steal" resources from future generations.[21] The United Nations has identified 17 sustainable development goals to transform toward a more sustainable future.[22] These goals guide the funding decisions of many national and international development agencies. These are but a few of many powerful global-scale actors working to create a more sustainable future.[23]

Reconnecting People and Nature

People's interactions with the rest of nature are sometimes destructive. According to Ovid's poems rooted in Greek legends, Phaethon asked his father, the sun god Helios, to allow him to drive the sun chariot across the sky for a day.[24] Helios refused because the horses were undisciplined, and the chariot was fiery hot. But Phaethon insisted.

When the day came, Phaethon was unable to control the horses. They veered from their course, scorching the earth, burning vegetation, changing much of Africa into desert, drying up rivers and lakes, and shrinking the sea. Earth cried out for help to Jupiter, who intervened by killing Phaethon with a lightning bolt.

I doubt the ancient Greeks ever imagined that mere mortals would enact this legend over ensuing centuries, leaving a human imprint of such magnitude and extent that geologists now realize that Earth has entered a new geologic interval (the **Anthropocene**) dominated by human forces that shape its future path.[25] The challenge is to identify potential future pathways more thoughtfully and to shape them wisely.

Homo sapiens is an amazingly adaptable species. Throughout the history of our species, we have responded to and shaped changes in our environment. Human beings were hunters and gatherers for hundreds of thousands of years. Then, 10,000 years ago, at the end of the last ice age, the climate warmed, and temperatures became remarkably stable (Figure 1.1).[26] This stability contributed to people's ability to cultivate certain plants and animals, especially those that yielded more or better food or were easier to collect. At the same time these precursors of domesticated plants and animals molded the behavior of society in ways that ensured their own success. Together, the domestication of plants and animals, along with other social processes, initiated a cycle of mutual dependence—in short, the birth of agriculture.

Around 1750, the Industrial Revolution began to harness energy from fossil fuels, which greatly expanded the physical capacity of the human enterprise. Industrialization and other changes allowed human populations to increase and draw more deeply on natural resources in the land and sea.

The course of human history has not been a smooth ride. In some times and places, wars, epidemic diseases, economic collapse, and multi-decadal droughts caused massive suffering. Civilizations came and went. At other times, bursts of progress—in agriculture, medicine, and technology— substantially improved life expectancy and quality of life over vast areas.

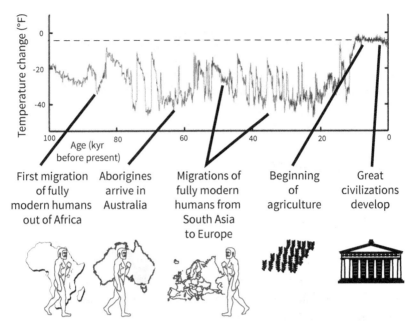

Figure 1.1 Variation in Earth's average temperature over the last 100,000 years, as recorded in glacier ice, and current understanding of the timing of important events in human history. Temperature is shown as the temperature difference from the instrumental record (1830–2000).

Even the good times had a dark side. As people cleared land and grew more food, the human population increased, so clearing more land for food was necessary. Each iteration of this cycle ratcheted up the need for food, minerals, fossil fuels, labor, and other resources, putting increasing pressure on Earth's lands and seas and producing a waste stream that spread pollutants far beyond their points of origin.[27]

So here we stand today. Is this just another eddy in the ebb and flow of history, in which things get better at some times and worse at others? Or has society set in motion processes and dynamics that push the planet toward a downhill slide that will be difficult or impossible to reverse? Today there are winners and losers, just as there always were. Turbulent change swirls around us, so society doesn't know with certainty what the future will bring. If these changes are "history as usual," perhaps society should just muddle along, addressing today's crises and opportunities and leaving future generations to figure out how to deal with the consequences of our actions.

But what if society's actions today condemn our grandchildren to live on a substantially less habitable planet with fewer options to thrive? Is Earth our planet to squander, or are we borrowing it briefly with the responsibility to sustain its capacity to nourish future generations?[28] Looking over our shoulder, we see that the once Fertile Crescent is now desert, the former forests of southern Europe are mostly scrublands, and the ocean has less fish. Are we destined to leave a similarly impoverished legacy over the entire planet?

The sheer magnitude and extent of human impacts on the planet over the past century are unprecedented. When global temperature first stabilized 10,000 years ago, there were perhaps 5 million people on Earth. Now, in 2020, we are 1,500 times more numerous—7.7 *billion* people. The human population is projected to increase by another 50% by the end of this century to 11 billion people.[29] That's about 200 people per square mile of land—including deserts, glaciers, and ice sheets—similar to the current population density of Virginia, North Carolina, and Europe. People in all of these places import food and other resources from elsewhere to sustain their populations. What will happen when there is no "elsewhere"?

In 1970, when Earth had half as many people as in 2018, our species annually used the equivalent of all the resources that Earth could regenerate in a year. In 2018, Earth's population annually used the equivalent of 1.7 planets of renewable resources.[30] But there is only one Earth, so how is this possible? Society is borrowing against the future—by clearing natural vegetation, depleting groundwater, overexploiting marine fisheries, and polluting the land, air, and water. In this way society reaps added benefits in the short run but reduces the capacity of Earth to meet the needs of future generations.

The question that keeps me awake at night is how can global society shape our planet's future so that our grandchildren and their grandchildren can thrive? Will future generations flourish with the legacy that society leaves them? Or will they ask why we chose to leave them a degraded planet with only photographs and stories of its former diversity and productivity?

In 2010, a drama group in Fairbanks, Alaska, performed a play about a family whose parents chose to maintain today's typical high-consumption lifestyle despite their awareness of its likely future consequences.[31] I still remember the lines of the now-adult children who confronted their parents in this play: "How could you have chosen a life that you knew would impoverish our world? Why didn't you care what would happen to us and our children?"

Greta Thunberg, a teenage Swedish climate activist, echoes this message: "Maybe my children and grandchildren will ask me why you, back in 2018, didn't do anything while there still was time to act. What we do or don't do right now will affect my entire life and the lives of my children and grandchildren. . . . We already have the facts and solutions. All we have to do is wake up and change. . . . We cannot solve the crisis until we treat it as a crisis."[32]

Society is now living out the lines of this play and is ignoring Greta's warning by consuming resources at rates that are unprecedented in the history of our species. If we continue down this path, most children born today will likely experience more warming during their lifetimes than has occurred since the dawn of human civilizations 10,000 years ago (Chapters 4 and 10). *Billions* of people would then find themselves living in a climate that today is found only in places like the central Sahara Desert.[33] The scale of the resulting human migration is difficult to imagine. Only by starting *now* to aggressively address the major causes of global change can these outcomes be avoided (Chapter 10).

Although scientists cannot *predict* the future, we know the factors that have contributed most strongly to recent changes. We could simply assume that these pressures will continue, as a result of more people, greater consumption of natural resources, innovations that continue to accelerate this development, and patterns of behavior that are similar to those of today.

However, the future might unfold differently. What if global society decided that the current diversity and productivity of the planet are worth sustaining for future generations? Is this outcome possible? After all, *Homo sapiens* is an ingenious and highly adaptable species. What are plausible pathways that might lead to such an outcome, and how might society foster these pathways?

As the late Catherine Attla pointed out, everything is connected, so the future depends on previous choices. In other words, the future is **path-dependent**. This dependence creates opportunities to shape the way the future may unfold.

I was first impressed with the importance of path dependence while watching the TV science fiction series *Voyagers*, when our kids were small. Our family would snuggle on the couch each week and watch with rapt attention as Phineas Bogg and a 14-year-old boy with a remarkable memory of history traveled back in time to give past events a nudge when history went awry. In one episode, Allied forces in World War I were being overwhelmed by German dirigibles because airplanes had, for some reason, failed to be invented. Our heroes traveled back to 1903, to find that the Wright brothers (in this story) had put aside their efforts to invent an airplane over competition for the hand of a beautiful woman. Phineas convinced the Wright brothers that inventing a flying machine was both important and feasible, and history then followed its observed course.

Unfortunately, in the real world, history cannot magically unravel itself and give society a second chance to get things right (or wrong), making it difficult to undo past environmental mishaps. Once a species is extinct, it's gone forever under current technology. The fossil-fuel CO_2 (**carbon dioxide**) already in the atmosphere commits us to centuries of continued human impact on climate.

Path dependence also means that today's actions can set in motion a set of *positive* changes that increase the likelihood of favorable outcomes, if we have a strategy to accomplish this. But setting out a pathway for positive change depends on whether we care about the future.

Both the causes and potential solutions to global problems are deeply rooted in human nature. Human beings are endowed with at least two primal instincts. One core human instinct is to compete for the resources needed to survive and reproduce, by hoarding those resources from others. This selfish instinct is shared with nearly every species on Earth and contributes to the success of competitive individuals.

A second instinct is to care for others and cooperate so that we may thrive together. This second urge, which is shared with many animals, is strongest for close relatives; but *Homo sapiens* often extends this care to the community and local environment on which each individual depends.[34] This second, gentler face of evolution contributes to the success of groups that share kinship, community, or cultural ties. This caring instinct has fostered the features of civilizations, cultures, and religions that make the human species remarkable and that we most cherish. Caring allows shared benefits to emerge.

Technology has expanded our species' capacity to compete with and care for others at scales never before imagined. Our competitive hoarding instincts contribute directly to global environmental degradation and widening disparities in power and wealth among social classes and nations. On the other hand, I'm optimistic that humanity's capacity to care provides both selfish and ethical motivations to fix some of today's most vexing environmental and social problems. In a world where nature is changing rapidly and natural disasters are on the rise, we care about others whom we may never meet. **Empathy** moves us beyond selfish instincts to care for kin and close community. It moves us to care for humanity, nature, and future generations.[35] In the process, society can create conditions in which people and the rest of nature can flourish together over the long run (Chapter 5).

I define **stewardship** as the process of shaping physical, biological, and social conditions to benefit both people and nature.[36] This word conveys a similar meaning to segments of society that sometimes fail to communicate their shared concerns. Stewardship is deeply rooted in Christian religious thought (care for God's Creation) and parallels concepts in other religions—providing common ground for conversations among faith groups about environmental issues.

Stewardship also captures the spirit of Aldo Leopold's writings in the 1940s about a land ethic of respect and care for nature. Stewardship has therefore become an important theme in modern conservation and resource management as a strategy to shape the health of ecosystems and society's long-term quality of life.

The scientific community was more hesitant to embrace the concept of stewardship. Some scientists thought that the word's religious and ethical connotations compromised science's claim to objectivity. However, as science expands its scope of inquiry to the Earth as a system, the massive human impacts on the planet have become obvious. The scientific community has therefore increasingly adopted sustainability science and stewardship as frameworks for applying science to the needs of society and the planet.

There has historically been a tension between religion and science, as well as between basic science and its application to society's needs. The accelerating pace of global change makes it increasingly clear that religion, science, and society must work together to address our shared concerns for our planet's future. Stewardship is one framework to foster this communication and collaboration.

The word "stewardship" traces its origins back to medieval England, where it referred to a guard (*weard*) who had a responsibility, on behalf of others, to care for a house or hall (*stig*).[37] In the timeless past, stewards cared for lands where lords and ladies hunted—perhaps the cool forests and mist-covered moors of the Scottish Highlands. Other stewards tended urban homes with garden paths, statues, and flowers that reflected the aspirations of their owners. In the urgent present, stewardship acknowledges that all people are members of nature's household and that each of us therefore bears responsibility for its care. Stewardship is an issue that needs us all.

I've never met a perfect planetary steward and doubt that I ever will. This book is intended for the rest of us, who want to make a positive contribution to Earth's future without completely sacrificing other things we care about. I will outline tangible steps that would enable any person to choose stewardship actions that fit their particular background, skills, interests, passions, and level of commitment to environmental and social issues. To illustrate differences in stewardship roles, this book emphasizes potential actions by four types of citizens: managers of private and public lands, indigenous people with strong cultural connections to their lands, city residents who encounter and shape nature in their neighborhoods, and people who tinker with nature in small gardens and woodlots or volunteer for community or conservation efforts. Each of these groups has, in its own distinct way, the experience needed to heal the land and the political power to trigger transformation.

People who make their living by working the land and sea, including farmers, ranchers, foresters, wildlife managers, and fishers, are particularly well poised to be effective ecosystem stewards. They often have the experience, intuition, and responsibility to manage lands and waters for sustainable harvest of products that society needs. They manage about half of Earth's terrestrial surface and much of the ocean and therefore have extensive influence over the future of our planet.[38]

Indigenous peoples inhabit an additional 25% of the global land area, often in remote places.[39] Their traditional cultural and livelihood relationships to their lands bring a long-term lens to stewardship.

More than half of the world's population now lives in cities, although cities occupy less than 3% of the world's land area.[40] City residents are therefore a major driver of human resource consumption and exert a large political clout in deciding the planet's future.

Individual citizens who tend small gardens and woodlots or volunteer in parks or wilderness inadvertently experiment with ways to learn how society can interact more effectively with nature. By discovering new ways to do things, they can potentially increase **resilience** (envision new potential response options) to the novel changes that are occurring.[41] These individual efforts also engage people to understand, celebrate, and share their experiences with nature through garden clubs, ranchers' associations, agricultural extension programs, and neighborhood gossip.

A Pause for Context

Before continuing, I will introduce you to the main characters of this book. As in any complex story, these characters are multifaceted, so you may associate them with different traits than I do. After briefly describing the identity and interrelationships among these characters from my perspective, we can be on the same page as we explore together their past and future lives.

Nature is the living skin of planet Earth. It includes all of Earth's organisms and the water, air, soil, and rocks with which they interact. People are part of nature.[42] We eat, breathe, and interact with the rest of nature, just as does any other species. The **biosphere** is all of Earth's nature—all of its ecosystems. **Ecosystems** consist of the interacting components of nature at scales ranging from a termite's intestine to the Amazon basin.[43] **Ecosystem health** is a metaphor that describes an ecosystem's condition, relative to its condition without negative human impacts.

Human beings (or **people** in ordinary words) are members of the species *Homo sapiens*. **Society** is a group of people that interacts at scales ranging from a neighborhood to an assembly of nations.[44] Society influences nature through individual and collective actions, regardless of motivation. **Humanity** is the collective human species, whose interactions with nature reflect a diversity of cultures and values. **Well-being** (or **quality of life**) is the health, happiness, and self-satisfaction of an individual or group of people.[45]

There is a tension between my human-inclusive view of nature and more romantic views, in which people are invisible or affect nature with a very soft touch. All species, including humans, interact with the rest of nature; but the human species deserves particular attention because of the magnitude of its impact and vested interest in its own long-term well-being.

Society and the rest of nature are linked by their effects on one another. **Ecosystem services** are nature's benefits to society.[46] These include harvested food, fiber, and water; regulation of climate, water quality, and disturbance; and people's cultural, spiritual, and recreational connections to lands and waters. **Environmental hazards** harm people through environmental stresses, ecological collapse, or other natural disasters.

Society's impacts on nature range from degradation to stewardship. **Ecosystem degradation** is the reduction in ecosystem health resulting from direct and indirect human impacts. **Ecosystem stewardship** involves human actions that shape Earth's future to restore, maintain, or enhance ecosystem health and human well-being.[47] Stewardship fosters resilience, if it increases the opportunities and options for responding to uncertain future change. The net effects of degradation and stewardship govern the **sustainability** of people's relationship with nature—drawing no more from nature than it can supply over the long run.[48] Degradation and stewardship are the actions or processes,[49] and sustainability (or lack of it) is the outcome. Linked social-ecological systems are sustainable if they meet current human needs without compromising the ability of future generations to meet their needs.[50] However, sustainability is more than just holding our ground. It implies outcomes that reduce hunger and poverty; improve access to healthcare, family planning, and education; increase agricultural production; and reduce environmental degradation.[51]

Currently, the net effects of stewardship and degradation are a mixed bag and vary from time to time and place to place. As endpoints, I consider outcomes that fail to meet sustainability goals as **impoverished** and those that meet these goals as **thriving**.

My greatest hope is that society can move beyond a materialistic desire to thrive—that we explore opportunities to **flourish**—to imagine with exhilaration and creativity what humanity could do and invent new ways to explore these aspirations. I want to watch people help their grandkids build a treehouse with magnificent views of sunsets and auroras, drink water from a clear stream rather than a plastic bottle, tend urban gardens so pretty that kids stop their parents to point them out, elect a mayor who believes more in bicycle paths than freeways, thank a politician who has funded societally meaningful jobs in places where people have given up hope, connect people who seek to protect their village from climate change with others who have found unexpected solutions, create a world where polar bears and pandas can thrive—and a million other things. There is no formula for flourishing. However, through stewardship for the health of both people and nature, society can create conditions that foster more widespread happiness, empowerment, and creativity. Individual and collective inspiration, imagination, and action then allow a culture of flourishing to emerge.

What Can We Do?

The first step in any journey is to decide where to go and how we hope to get there. This helps us understand and explain to others why we have chosen our particular path. As individuals, our best options for contributing to society's and nature's opportunities to flourish include the following:

- **Be informed**. Know, in general terms, the historical reasons why human interactions with nature have deteriorated. The references in the back of this book taught me a lot about these issues.
- **Know what works**. Point to positive examples of ways that society has improved those human-nature interactions that we, as individuals, most care about.
- **Imagine a better world**. Think how people could interact more effectively with nature so that both may flourish.
- **Take responsibility**. Commit to taking responsibility for some actions, no matter how large or small, that will improve society's interactions with nature for the benefit of both.

The next question is: *Given the current trend of environmental degradation, do we know enough to shape the future wisely?*

2
Sustaining Natural Change

Scientists understand the general conditions that enable ecosystems to thrive over the long run. This chapter describes these foundations of ecosystem sustainability and ways that society can foster these conditions.

Taming Hubris

People have interacted with their environment throughout human history. This began at very local scales with hunter-gatherers, fishers, and farmers. As human population increased, people transformed extensive areas into farms, second-growth forests, grazing lands, and cities. This altered their relationships with the rest of nature. People were optimistic that they could tame nature to serve human purposes. Specialists were trained to manage many of the human uses of the land and sea. However, new land uses sometimes had unintended consequences. As transformed lands became more extensive and interconnected, for example, pest outbreaks and diseases that arose in one place spread more easily to adjoining lands.[1] Does the goal of shaping the future health of ecosystems reflect **hubris**—overconfidence in human capacity to plan for the future?

Unfortunately, resource management has a checkered history of meeting many of its *short-term* goals. Australia's first European colonists brought sugar cane with them as a potential export crop.[2] It grew well in the warm, wet climate of northeast Australia, so sugar plantations thrived and spread. By the early 20th century, sugar was a major export crop. Sugar cane also served as food for a native Australian beetle, whose larvae fed on cane roots and periodically devastated the crop. At about the same time, cane toads, the largest species of the frog family, seemed successful at controlling cane beetles when introduced to Puerto Rico from tropical America. Many sugar-producing regions of the world, including Australia, therefore imported cane toads in an effort to solve their pest problems. Unfortunately, in Australia, the cane toads preferred natural habitats and foods to the cane fields. The toads expanded rapidly through northeast Australian ecosystems without having any detectable impact on the cane beetle. The toad's poison glands are toxic to predators, leading to large declines in many native Australian predators, such as monitor lizards. They have become a major conservation threat, while failing to provide relief to the sugar industry. Cautious small-scale experiments could have easily tested for some of these unintended effects of cane toads before a full-scale introduction was launched.

The cane-toad disaster is symptomatic of problems that occur when managers try to solve immediate management problems without carefully examining a broad range of ecosystem effects that a potential introduction might trigger. Species that are moved to new places can become superabundant if they escape predators or diseases that held their populations in check back home. Rabbits that were brought to Australia to be raised for meat bred—well, like rabbits—and quickly became so abundant that they degraded extensive areas of Australian grasslands. Only when a rabbit viral disease was introduced were rabbit populations reduced to reasonable numbers. Rabbits then adapted to the disease and partially regained their previous numbers. Red deer, opossums, and weasels were introduced to New Zealand, a country whose mammal fauna prior to human arrival included only bats and sea mammals.[3] These and other introduced mammals outcompeted or ate many of New Zealand's native ground-nesting birds, leading to a pulse of species extinction. Similarly, pines in South Africa and European grasses in California now dominate their new homes—and the list goes on.

Attempts to develop agriculture without considering the local ecology can also lead to bad outcomes. The Soviet Union tried to develop agriculture near the Aral Sea in the 1960s (Figure 2.1).[4] Land planners diverted water for irrigation from the rivers that fed what was once the fourth largest lake in the world. Reminiscent of Phaethon's fateful journey across the sky, the Aral Sea shrank to 10% of its original size, the dry former lakebed became a desert littered with rusty fishing boats, and 90% of the people left the region. Salty sands from the dry lakebed blew across irrigated fields, reducing agricultural production; and temperatures soared as the cooling influence of the lake was lost.

As in the cane-toad and Aral-Sea examples, managers who pursue short-term goals without understanding or considering broader contexts often encounter unexpected outcomes that can negatively affect both ecosystems and local residents. However, society can learn from these mistakes and build on the wisdom that comes from deeper understanding of ecosystems.[5]

Figure 2.1 Aral Sea in 1989 and 2014 (top). Fishing boats stranded when the former lakebed of the Aral Sea changed to desert (bottom).

The Goldilocks Principle

Someone who knows a forest well knows intuitively where it falls on a spectrum from degradation to health and can confirm his or her intuition with measurements and cautious experiments. When I fly across Alaska's landscapes, I have a comfortable sense of familiarity—a manageable set of ecosystem types that stretch from horizon to horizon. As an Alaskan ecologist, I have come to know these few types quite well—needle-leaved forests, shrubby muskeg, tussock tundra, and a few others. But it takes time.

Only gradually did I begin to develop an intuitive sense of why each ecosystem type occurs where it does and how it works.[6] To know a black spruce forest well, I must rub shoulders with scrawny centenarian trees whose trunks are the thickness of my arm and walk over the spongy moss carpet that reluctantly shares its sparse nutrient supply with the spruce. As I lie on my belly with my nose in the moss to document the flow of nutrients that support moss and tree growth, I get a closer sense of it all: the musty smell of decomposing earth, the slightly out-of-tune whine of mosquitoes, the white fungal networks that weave through a miniature Manhattan of mosses to connect recently shed spruce needles to moss fronds and fungi-encrusted tree roots beneath. Reaching down through the cold, wet peat beneath the mosses, I feel the **permafrost**—the frozen soil that keeps water perched near the surface, where it dictates everything that happens above.

By measuring the environment and quantities and flows of carbon and nutrients in this forest, scientists like me translate these sensory perceptions into numbers that make sense to other scientists and are important to society. As permafrost thaws due to recent warming, for example, it releases carbon that has been stored for thousands of years. The quantities released to the atmosphere as carbon dioxide (CO_2) in today's warmer climate are now similar to amounts emitted globally by society through the burning of fossil fuels.[7] Climate warming caused by human actions has unleashed a new source of CO_2 to the atmosphere that amplifies human impacts on the climate!

After living with this forest for five decades, I know many of its quirks and habits and why it has chosen to exist in this particular place. I also learn how this ecosystem might behave differently if its environment changes. Adjacent to the black-spruce forests are warmer south-facing slopes covered by white-spruce and broadleaved birch and aspen forests. These forests may be a preview of areas now occupied by black-spruce forests if climate continues to warm. In cool years, such as those that shaped the boreal forest over the last 6,000 years, wildfires leave behind a thick layer of charred peat, in which only black spruce can germinate and grow, ensuring its persistence in the regenerating forest. However, my colleagues Jill Johnstone and Teresa Hollingsworth find that, in the unusually warm years that occur more often in recent decades, wildfires burn away most of the moss and peat. This leaves behind a moist mineral-soil surface on which seedlings of other trees, especially the more rapidly growing birch and aspen trees, establish, grow, and eventually dominate the forest.[8]

Based on satellite images, Scott Rupp and Dan Mann discovered that Interior Alaska has shifted in the last half-century from dominance by needle-leaved to broadleaved forests.[9] From pollen profiles recorded in lake sediments, Linda Brubaker, Andrea Lloyd, and other paleoecologists know that this is the first time in 6,000 years that such a widespread vegetation change has occurred.[10] Perhaps this recent change is not surprising since northern ecosystems, like those in Alaska, are warming twice as fast as the average for the planet.[11] These changes in forest composition will probably alter vegetation effects on climate, fire probability, and productivity and therefore the effects of these ecosystems on society.

Thousands of ecologists and people who have spent their lives working the land and sea have an intimate understanding of their ecosystems and know the conditions and interventions that are likely to either sustain current ecosystems or change them to something new. This ecological understanding underpins predictions of how these ecosystems, including managed lands, may change if the world continues to warm.

California's astounding ecological diversity is a daunting contrast to the comfortable monotony of broad expanses of a few major vegetation types in Alaska. My favorite teaching gig as a faculty member at the University of California Berkeley was a field course in ecosystems of California. Here I was only a few steps ahead—and often behind—my students in trying to sleuth out why the plants, animals, and soils in one place differed so strikingly from those in the next valley.

On one family outing—which also served as a dry run for my next class field trip—my wife Mimi and I visited the towering redwoods of Muir Woods. We drove through eucalypt forests, annual grasslands, and scrubby chaparral before suddenly dipping into cool redwood groves—then back to a mix of grassland and chaparral. These patterns were a real head-scratcher for someone like me who lacked the intuition that comes with long familiarity. After visiting and reading authoritative articles about each California ecosystem, I often had just as many questions as beforehand. Therefore, before each class field trip, I would talk with people who knew their ecosystem well, to see if I could glean some intuition from their experience.

Hans Jenny was one such giant from whom I learned a lot. He was 90 years old and about 5 feet tall when Mimi and I drove with him to his most famous field site, the ecological staircase in Mendocino County of northern California. He had just begun a new field experiment and was glad for the ride so that he could see how things were going.

Hans Jenny was one of the founding fathers of ecosystem ecology. By his reckoning, scientists can predict the properties of any ecosystem—especially the soil, which was his particular interest—based on a few independent factors that shape its properties.[12] This recipe includes the types of rocks that supply soils with nutrients, topography, climate, regional flora and fauna, and time. He later added human activities to his list of critical ingredients. This formula is important because it provides a framework for understanding why every ecosystem occurs where it does and why and how it might change in the future.

Hans tested his idea about the role of independent factors in controlling the properties of ecosystems by comparing places that differed in only one of these factors. Mendocino County was one such place.[13] Over the last half-million years, periods of uplift along the California coast had created a series of marine terraces or "stairsteps" as land emerged from the sea. Each uplift interval added a new step to the staircase. Each of these terraces had virtually identical topography, rock substrate, climate, regional suite of organisms, and absence of direct human disturbance but differed in the time during which soils and vegetation had developed.

Hans took us to his favorite spot on each step of his staircase. Starting at the youngest stair, closest to the ocean, we looked down from the beach-cliff edge. The flotsam of seaweed was the only evidence of plant life on the wave-battered sand below.

Hans explained how soils developed through time. On the terrace above the cliffs was a grassy herb-rich meadow with sandy soils that were too dry in summer to support trees. On the next-oldest step, some of the sand had broken down into finer particles, which, together with accumulating organic matter, retained enough water and nutrients to support a redwood forest with occasional Douglas firs and bishop pines (Figure 2.2). On the final terrace was a pygmy forest of pine and cypress trees only slightly taller than ourselves. On this terrace, the acidity produced by the decomposing leaf litter had broken down soil minerals, which leached downward to produce an impermeable soil layer (**hardpan**). The soil above the hardpan, in which plants were rooted, was nutrient-impoverished and often waterlogged, providing very poor conditions for tree growth.

When we delivered Hans back to his Berkeley home, he showed us his paintings of soils, illustrating the beauty and diversity of this hidden foundation that nourishes ecosystems to varying degrees when nature is allowed to take its course. He clearly felt a deep personal connection to his subject.

On the ecological staircase, Hans had selected sites where terrace age was the major differentiating factor. If he was correct, the same principles should apply to the distribution of all ecosystems on Earth. Each ecosystem occurs in a Goldilocks relationship with its environment, where its particular

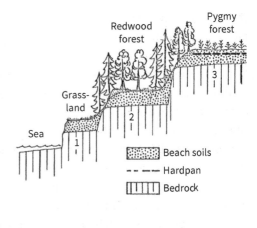

Figure 2.2 Hans Jenny examining a soil profile in a nutrient-rich redwood forest (left). On the right is a modification of his diagram showing three marine terraces, each with distinct soils and vegetation.

requirements are matched with the local soil, topography, climate, organisms, age, and disturbance conditions—just as Goldilocks chose a bowl of porridge and a bed that were just right for her.[14]

Productive redwood forests, for example, grow in stream valleys and lower slopes where deep soils provide abundant water and nutrients and where tall ridges hold the fog that shields tree canopies from the desiccating summer sun. Scrubby chaparral, on the other hand, occurs on drier upper slopes, where erosion-prone shallow soils provide less water and nutrients and the sun and wind suck away most available moisture. Both ecosystems have a productivity that matches the soils and climate where they occur. These interactive relationships between plant species, soils, and climate explain the fine-scale landscape patterning that I found so confusing when I first moved to California.

This description is, of course, an oversimplification because history matters, and changes in climate, disturbance, and species can alter both the composition and functioning of ecosystems. When these changes are modest, landscapes adjust and sustain similar species composition, structure, and process rates for thousands of years. But now the environment is changing more rapidly than many ecosystems can adjust, raising questions about how to sustain or restore their health.

Managing Sustainable Change

Sustainability is more than keeping things the same. Given today's rapidly changing environment, how do we shape changes in ways that sustain the health of ecosystems and the well-being of society so that both may flourish?

When early Euro-American pioneers moved west, they found prairies with deep soils that seemed ideal for agriculture. By the late 1800s, they had plows capable of converting large areas of native prairie to cropland. During the epic drought of the 1930s, soils dried out on both croplands and overgrazed grasslands.[15] Up to 75% of the soil that had accumulated over thousands of years blew away in massive dust storms in less than a decade. Dust blew eastward as far as Boston and New York, and the people who could no longer survive on their impoverished lands moved westward as far as California (Figure 2.3). When rains returned to the Dust Bowl in the 1940s, there was less soil and therefore less capacity of soils to store water and nutrients to support crop growth. Rain-fed agriculture therefore became permanently more drought-sensitive and less productive.

Soil loss in the 1930s Dust Bowl shows how changes in a key environmental determinant can permanently impoverish ecosystems. When erosion removes soil faster than it is replenished by the slow breakdown of rocks and other inputs, the soils hold less water and nutrients to support plant growth; and therefore, the ecosystem adjusts—in Goldilocks fashion—as less productive plants adapted to these conditions replace those that were there before.

Soils can recover, just as Hans Jenny showed in his ecological staircase; but recovery takes a long time. This process can sometimes be speeded by planting species that are especially effective at replenishing soil fertility.[16] Native Americans and small-scale farmers have discovered many ways to sustain and restore soil fertility, for example, by intermixing nitrogen-fixing crops like beans with those that require more nitrogen. **Nitrogen-fixing plants** host bacteria that convert atmospheric nitrogen to forms that plants can absorb. Researchers in sustainable agriculture have developed many strategies, often learned from nature or from small-scale gardeners and farmers, for sustaining soil fertility by retaining, recycling, or replenishing organic matter and nutrients.[17] These approaches reduce the need for energy-intensive synthetic fertilizers and provide many additional ecological and health benefits (Chapter 3).

Figure 2.3 Dust storm in Stratford, Texas, 1935 (top), and Oklahoma farmers who broke down on a highway near Lordsburg, New Mexico, in 1937, after being forced westward by drought and soil loss in the Dust Bowl (bottom).

Maintaining soil quantity and its contents of organic matter and nutrients are perhaps the easiest ways to maximize the productive potential of highly managed ecosystems. It's always easier and more effective to minimize nutrient and soil loss in the first place than to fix problems that result from excessive erosion. The muddy nutrient-rich rivers flowing from many farmlands show that soil erosion and nutrient loss are still widespread problems.

Early Spanish missionaries, military, and traders accidentally brought many annual plant species to California from heavily grazed grasslands back home.[18]

At first these plants were probably restricted to disturbed areas around newly built missions and presidios. California and Spain shared a Mediterranean climate, so the new arrivals prospered in disturbed sites where there was space for them to get a foothold.

The California Gold Rush of 1848 changed everything. The Euro-American population quadrupled in a few years, and farmers stocked extensive areas of California grasslands with cattle or converted them to agriculture to feed the influx of people. Continuous intensive cattle grazing and associated soil disturbance replaced seasonal elk grazing and periodic burning by Native Americans. These and other changes enabled European annual grasses to outcompete their perennial California counterparts over extensive areas. Once the European species dominated, the native plants had difficulty re-establishing, so most California grasslands have been dominated by European annual plants for more than a century. Annual grasses lack the deep roots that enable California perennials to access late-season moisture, so the productivity of annuals is more sensitive to yearly variations in weather. This shift in plant species changed many of the ways in which California grasslands function.

Just as in California, nature is constantly rearranging itself. When land is cleared or new species arrive, the new assemblage of species usually differs from what was there before. The changes that emerge depend on the species that are present locally and their capacity to grow in their new home. Scottish heaths became "natural vegetation" after primeval forests were cleared. Many "wilderness" areas in US national parks are still changing in response to the removal of Native Americans who managed these lands 150 years ago. Old-growth forests of eastern North America lost their once-dominant chestnut tree after a fungus accidentally escaped from Japanese nursery stock in 1904. The fungus eliminated chestnut trees across the eastern United States.[19] However, all of these places show the exuberant resilience of nature's capacity to flourish. Although things have changed, society can still benefit from and celebrate the human-shaped nature that develops around us, even though we might prefer that these changes had never occurred.

As Aldo Leopold pointed out, the intelligent tinkerer keeps all the parts because we never know when they will be needed.[20] In this context, sustaining the current properties of ecosystems and conserving lots of species give future society more options (greater resilience) to fix the problems we leave for them. This is most easily done by reducing novel pressures that are causing current ecological change—in other words, sustaining Hans Jenny's critical factors of climate, soils, organisms, and patterns of disturbance.

Keeping an ecosystem's current species is challenging because new species are accidentally introduced from distant places in ever greater numbers. It would be helpful to know which kinds of species are most likely to cause large changes in ecosystems. In this way, precautions can be taken to minimize the likelihood of their arrival and establishment. Novel species that are most likely to disrupt ecosystems are those that alter (1) the feeding relationships among species (species like top-level predators or diseases), (2) disturbance patterns (highly flammable species), (3) water and nutrient availability (nitrogen-fixing or deep-rooted species), and (4) ecosystem structure (trees in a grass-land).[21] Novel species interact with the rest of nature in so many complex ways that the arrival of a new species often leads to unexpected outcomes—as with cane toads in Australia.

In many cases we care about a particular species and want to protect it. Connie Millar, a forest geneticist, was concerned about the impacts of climate change on an important forestry species (Douglas fir) in the Sierra Nevada of California.[22] Well-established forestry practices dictated that logged forests should be replanted with locally collected seeds. This practice would sustain the local genetic composition of the forest in its Goldilocks relationship with local conditions. However, this strategy would doom trees to severe environmental stress if the climate changed over the 75–200 years required for trees to reach harvestable size. Connie chose a strategy that she learned from nature's evolution. She planted a range of genetic types that were adapted to a variety of current and potential future climates. She didn't decide which genotype was best—she let nature sort out the types that would survive and prosper. This is one of a toolbox of approaches that she and her colleagues explored to foster forest resilience to a rapidly changing environment. Her work illustrates the ingenuity that must be learned from nature and tested cautiously to shape the future health of ecosystems.

The impact of climate change on species migration is especially controversial. Species have always migrated to stay in tune with their Goldilocks climate, but what should managers do if species cannot migrate fast enough to keep pace with climate change? Should species be moved to places where conditions now mimic their historical climate (**assisted migration**), or should they be left to battle for survival in a climate that becomes increasingly unfavorable in their current locations?[23]

Saguaro cactus and Joshua trees, iconic symbols of southwestern desert vegetation, are now trapped in an increasingly alien environment (Figure 2.4). A combination of warming temperatures, nitrogen pollution, and invasive European grasses has substantially increased fire risk to fire-vulnerable saguaro and Joshua trees in the heart of their ranges.[24] Other locations that

Figure 2.4 Saguaro cactus in the Sonora Desert (top) and Joshua trees in Joshua Tree National Park (bottom).

are cooler and less polluted have no invasive grasses and would appear to be good locations for saguaro and Joshua trees if climate continues to warm. However, seed dispersal is so slow and seedling establishment so infrequent that these iconic species have been unable to reach these suitable habitats.

Similarly, frogs that are restricted to wet places and pikas that inhabit rocky mountain tops face migration barriers of inhospitable habitats, as do some animals that encounter an eight-lane highway. Should managers give these species a helping hand to reach their Goldilocks climate? The safest options are probably to reduce the rate of climate warming (Chapter 4) and to maintain migration corridors of relatively undisturbed habitat that would allow at least some species to migrate at their own pace as best they can.

Sometimes people care more about what a species does than what it is. Bees are a pragmatic example of the usefulness of understanding the functional properties of species. Pollinating insects are essential for the reproduction of many flowering plants. Social bees, such as honey bees, nest together in hives, where they share information about the location of pollen sources. This makes them very efficient at finding and pollinating flowers. This social habit also makes social bees vulnerable to insecticides and pathogens that foraging bees accidentally bring to the hive. This insecticide vulnerability has contributed to worldwide declines in many species of social bees.[25] Parts of southwest China are so polluted by insecticides that there are no bees to pollinate fruit orchards.[26] Here, China has resorted to labor-intensive hand pollination of orchards by people to sustain fruit harvests (Figure 2.5). Although insecticides are intended to reduce insect pest populations, their unintended impacts on pollinating insects or

Figure 2.5 Hand pollination of fruit orchards in China.

on predators of insect pests sometimes have net detrimental impacts on agriculture.

Disturbances such as fires and floods are natural components of all ecosystems. Most landscapes are therefore patchworks that reflect a history of many types and sizes of disturbance.[27] The impact of disturbance is obvious in landscapes with a heavy human imprint. Even in "natural" ecosystems, however, gopher mounds, treefalls, avalanches, floods, or wildfires create landscape patchworks. Unfortunately, people often suffer from disturbances that are an inherent part of nature. Society therefore often tries to prevent fires or floods despite their necessity for sustaining normal ecological processes in fire- and flood-dependent ecosystems.

Perhaps the most resilient solution is to match the human use of ecosystems with the disturbance patterns dictated by changing patterns of climate and ecology. Climate warming has led to more severe drought and fires in dry regions, extreme rain events in wet regions, and increased frequency of storm surges along coasts (Chapter 4). The societal impacts of these changes in disturbance regime can be minimized by concentrating development in places with a low risk of life-threatening disturbances. However, this didn't happen in Houston, Texas.

Houston grew rapidly in the 1990s and early 2000s, fueled by an expanding oil and gas industry. Planners and developers paid little attention to international trends of increasing frequency of intense hurricanes. Many of Houston's new developments were built in low-lying areas, where open land was still available. In 2017, Hurricane Harvey flooded many of these areas, putting 30% of Harris County under water.[28] Development made the flooding more extreme in two ways. Much of the development occurred on drained wetlands that would otherwise have temporarily retained and absorbed floodwaters. In addition, the large area of impervious surfaces, such as buildings, roads, parking lots, and malls, rapidly delivered water to streams and rivers that flooded downstream areas.

The flooding of Houston during Hurricane Harvey was not entirely surprising. Houston's flat topography and location on the hurricane belt makes it one of the most hurricane-vulnerable cities in the United States. Deaths from severe flooding had occurred in Houston in both of the two years immediately before Hurricane Harvey. Much of the flood damage in Houston could have been averted by protecting natural wetlands and other low-lying areas from development. Now, after Hurricane Harvey, Houston continues to develop low-lying areas—a recipe for almost certain future disasters.

Coastal houses face analogous risks from rising sea level. These risks can be minimized by restricting development in locations that are becoming more disturbance-prone. A more common outcome is for some people to rebuild in the same place, when their home is destroyed. Others move away to safer places after catastrophic coastal disturbances such as Hurricane Katrina in New Orleans and Superstorm Sandy near New York City.[29]

Some communities that recognize their increasing vulnerability to disturbance proactively shape landscapes to increase their resilience to these risks. Phoenix, Arizona, and Cardiff, Wales, have city parks adjacent to flood-prone rivers. Most of the time these parks are sites for recreation and family picnics. These amenities make adjoining neighborhoods attractive places to live and attract new homeowners and businesses to the area. When big floods do occur, the parks store floodwaters and release them slowly as the flood recedes.

Similarly, two nongovernmental organizations (NGOs), Headwater Economics and Wildfire Planning International, have teamed up to provide training, recommendations, and support for land use–planning tools that reduce wildfire risk to communities in fire-prone regions.[30] One of their strategies is to prioritize agriculture, golf courses, and other open-land uses as buffers around communities. They then concentrate development inside these fuel-free buffers—analogous to medieval European towns where homes and shops were concentrated inside fortifications that protected them from marauding enemies. It would be equally logical to restrict human development in highly flammable forests and shrublands that cannot be fully protected from wildfire. This can be done through zoning that restricts development in high-risk areas or compulsory fire or flood insurance that requires individuals to assume the economic risk of building in unsafe places.

When Mimi and I moved back to Alaska from California, we bought a house in the woods. As an ecologist who studies Alaskan wildfires, I knew that the broadleaved trees around this house were much less flammable than spruce, so we cut the few spruce trees that were close to the house and trusted nature's protection against wildfire by broadleaved trees. In the 20 years since we moved, however, the warming climate has changed the ecology of wildfire. During extremely dry weather, which now occurs more often, any tree will burn, so the landscape has become more flammable; and we worry when the increasingly frequent wildfire smoke darkens the sky. Many people now find their lives and homes more vulnerable to disturbances than they expected. Collective action, like that recommended by Headwater Economics, may be the best option for minimizing the risks of changing climate for future development.

Implementing Stewardship

Traditionally, stewardship involved management by stewards of the ecosystems under their care, without much thought about the interactions between people and nature (Chapter 1). This narrow view of stewardship is similar to current top-down management of many parks and private lands based on the preferences, rules, and regulations that guide the managers of these lands. This traditional form of stewardship can meet many of the ecological goals of stewardship, when it is based on sound ecological principles. However, top-down management usually reflects the priorities of managers, rather than those of society as a whole. For example, forest lands may be managed primarily for timber harvest or fire control, rangelands for cattle production, and parks for aesthetic or cultural experiences of visitors. There is no guarantee that narrowly defined *ecological* stewardship will support the suite of values and concerns of a diverse society.

Europe's long history of intense human interaction with the land created a rich spectrum of experience in land uses ranging from working landscapes to those where people play a smaller role. This became problematic in the 1960s, when France established its first national parks. In these new parks, policies protected nature from human activities in a core park that was surrounded by a peripheral zone where social, economic, and cultural development was intended to be "compatible with biodiversity conservation, rural life, and local culture. It should also attract tourists in search of natural and cultural landscapes and local traditions."[31] However, people in peripheral zones resented the new restrictions imposed in the protected core areas. The French national park service was also frustrated that it could not compel people to implement the changes it wanted in peripheral zones.

In 2006, France redesigned its national park system according to a new principle of **ecological solidarity**, which meant that decisions were intended to benefit all **stakeholders** (people and other species that are affected by an organization's decisions), rather than privileging just nature or just visitors. Communities near national parks were invited to designate some of their territory as part of the park by adhering to the park's charter.[32] The new park charter actively engaged community members in managing peripheral areas in solidarity with the park's conservation objectives and with the community's cultural, social, and livelihood goals. Ecological solidarity provides principles that guide *social-ecological stewardship* rather than more narrowly defined *ecological stewardship*. Ecological solidarity emphasizes the importance of effective dialogue and collaboration among stakeholders (Chapters 7 and 8) and sharing power to make decisions together (Chapter 8). It's an opportunity for local residents to have a voice in the lands they care about. It builds on people's empathy for nature and society at large.

Social-ecological stewardship is being increasingly applied to complex landscapes throughout the world in ways that support a diversity of land uses and the values and identities of people. Not surprisingly, the degree to which management actually implements social-ecological stewardship varies from place to place, depending on local culture, politics, and environment.[33] The European Union's Natura 2000 policy fully engages local landowners, communities, NGOs, and parks in collaborative management on about 15% of Natura 2000 sites.[34] It collaboratively guides the management of Europe's protected areas. That's a good start but not yet a fully implemented solution.

What makes some stewardship efforts more successful than others? In their project "Seeds of a Good Anthropocene," Elena Bennett and her colleagues examined 100 sites around the world where social-ecological stewardship appeared to have flourished.[35] These ranged from rural food-producing systems to cities and were sparked by innovations drawn from many cultural traditions and academic disciplines. Most of these seeds began locally as bottom-up efforts led by local groups of citizens and produced outcomes that were good for both people and nature. The Satoyama Initiative in Japan, for example, connects urban and rural areas to reinvigorate traditional agricultural landscapes that were declining because of rural migration to cities. City visitors now provide labor on farms as they learn about traditional farming and culture. New York City's High Line Park made native species, art, education, and recreation accessible to all by repurposing an abandoned elevated freight rail line.

In addition to these and many other initiatives, millions of people regularly interact with their lands and waters as they pursue livelihoods and hobbies and cope with challenges of weather, pests, and disturbances. In this way, people regularly observe and experiment with what works and what doesn't as they adjust to change. As in Stockholm's allotment gardens (Chapter 1), friends share what they learn. The most helpful tips are shared broadly through associations of gardeners, ranchers, and farmers and are often publicized in news reports, blogs, and social media. Agricultural extension services and citizen science networks connect the learning networks of private citizens with those of scientists.

The Wolf Bay Watershed Watch is a nonprofit organization of trained citizen volunteers who monitor water quality of the Wolf Bay Estuary in coastal Alabama.[36] This estuary is home to several threatened species, including the red cockaded woodpecker and Florida manatee. Water quality of the habitat was threatened by mushrooming regional population. Rigorous water-quality monitoring by citizen scientists convinced the Alabama Department of Environmental Management to designate Wolf Bay as an Outstanding Water. This designation limited the allowable pollution discharge and helped residents learn why it was important to protect their bay. Citizen science made a difference.

Broad networks for learning and sharing are increasingly important in today's world because of rapid changes in those factors that govern ecosystem functioning—climate, organisms, disturbance regime, and soils. Effective learning and sharing networks make society more aware, knowledgeable, and resilient as we test and implement responses that appear adaptive and curtail practices that cause bad outcomes.

Phoenix, Arizona, increasingly encountered water shortages as population and water demand increased at the same time that warmer temperatures and greater evaporation reduced water availability. By charging residents for summer water, the city encouraged residents to plant **xeriscapes**—plants that require very little watering and are typical of the natural desert vegetation of the region. Two-thirds of homes in Phoenix shifted from heavily watered lawns to xeriscapes, and water use per person declined by 30% between 2000 and 2019.[37] In this way, citizens became more aware of the nature around them at the same time that they saved water.

Madison, Wisconsin, found that overfertilization and overwatering of lawns were polluting Lake Mendota, a prized recreational site near the city center. As the city publicized practices to reduce this pollution, citizens became more aware of ways that their actions were integral to the nature they cared about.

All types of citizens can play important roles in discovering ways that ecosystems can remain healthy and resilient, despite rapid change. Farmers, foresters, fishers, and resource managers know from personal experience the importance of being tuned into the changes around them, as do indigenous people who depend on their lands for hunting, fishing, and herding.

Tinkering by individual citizens through small-scale experiments is a relatively safe way to learn about ecological responses to unexpected changes. Every farmer has a favorite way to farm. Every gardener chooses different flowers or vegetables, which influence the pollinators and birds that visit their garden and city. All forest owners manage their forests differently. This diversity of approaches creates an ideal learning laboratory for improved understanding of ecosystem responses to novel conditions. Besides, these experiments are fun! Engaging people and implementing and sharing results of their experiments foster learning and resilience (Chapter 7). Parks, garden clubs, and agricultural extension agents usually jump at the chance to share success stories broadly.

Individual citizens are also well positioned to observe emerging problems as people walk, picnic, and volunteer in wilderness, parks, and gardens—too few bees or butterflies to pollinate the flowers in their garden, the arrival and spread of a new invasive species, widespread disease or death of trees that used to be healthy, superabundance of deer and ticks in suburban forests. Sounding the alarm bells about potential problems can help trigger searches for solutions before it's too late.

What Can We Do?

Stewardship actions to sustain ecosystems require people with diverse skills and relationships to the land, water, and sea. There is a role for everyone:

- **Observe and record local ecological changes.** People who work their lands or gardens, volunteer in parks or ecosystems, or spend time in wild or urban nature are well positioned to notice ecological changes.
- **Learn why ecosystems change.** Read, ponder, and learn how Hans Jenny's critical factors—especially climate, water, nutrients, pollution, organisms, disturbance, and human activities—are changing in ways that might affect the ecosystems we value.
- **Experiment cautiously.** People who manage land, water, and sea— farmers, ranchers, foresters, fishers, land managers, and local gardeners— can experiment cautiously to test which changes most strongly affect valued ecosystems and how people can intervene to shape ecosystem responses.
- **Share knowledge and insights.** Sharing knowledge learned from reading, observing, or tinkering with ecosystems builds a platform for coordinated public action. Consider joining a citizen science network.

The next question is: *If we know how to sustain nature, how can we ensure that both nature's and society's needs are met?*

3

Linking People with Nature

Linking people with nature is essential to the health of both. This chapter describes how to shape these linkages so that both people and nature can flourish.

Nature's Benefits to Society

Wangari Maathai, the daughter of a Kenyan sharecropper, collected water for her mother from springs protected by the roots of trees.[1] Her grandmother taught her that the fig tree near her home was sacred and should not be disturbed. Although it was unusual for girls to attend school, her brother suggested it, her parents agreed, and she graduated near the top of her class. She was the first woman in eastern and central Africa to receive a PhD. In the 1970s, when she was in her mid-30s, she became aware of Kenya's ecological decline and its link to rural poverty. Watersheds were drying up as native forests were cleared for farms or plantations of fast-growing, water-demanding exotic trees. "I was hearing many rural women complain that they didn't have firewood, that they didn't have enough water. So why not plant trees, I asked the women."[2] She thought that planting native trees, especially fruit trees, might replenish soils, provide fuel wood, and improve nutrition in rural communities.

Maathai began by raising tree seedlings in tin cans in her backyard and engaging rural women in planting them. Government foresters initially resisted the idea because they didn't believe uneducated rural women could plant and tend trees. Little by little, the idea took root and blossomed into the Green Belt Movement. Maathai showed women how they could use their existing knowledge to gather seeds from the forest, plant them, and tend the seedlings. Eventually, with help from Kenyan and United Nations groups for the support of women, she was able to pay women 10 cents for each tree that survived and grew. This gave the women a sense of independence and empowerment. The Green Belt Movement spread across East Africa, and the 51 million trees planted by these women have significantly reduced land degradation in the region.

When Maathai accepted the Nobel Peace Prize in 2004, she said that the purpose of her program was to help people "make the connections between their own personal actions and the problems they witness in their environment and society. . . . With this knowledge, they wake up—like looking in a new mirror—and can move beyond fear or inertia to action."[3] Clearly, even people with minimal material wealth and security can work together to do amazing things that simultaneously improve their lives and the environment.

As Wangari Maathai showed, neither ecosystem health nor human well-being is sustainable by itself because each depends on the other. People shape nature through exploitation, impacts, and stewardship; and nature shapes people through its benefits to society (ecosystem services) (Figure 3.1).[4] Analogous to the social and commercial services provided by government and business, ecosystem services include the following:

- **Provisioning services**: the products that are directly harvested from ecosystems (such as food, fiber, and water)

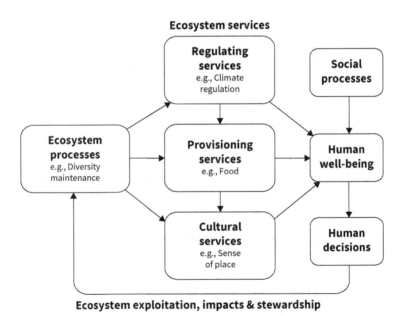

Figure 3.1 The relationships between the benefits that people receive from ecosystems (ecosystem services), human well-being, and the human actions (ecosystem exploitation, impacts, and stewardship) that influence ecosystems and their services to society.

- **Regulating services**: the capacity of ecosystems to buffer disturbance and shape interactions among ecosystems (regulating the climate, cleaning our drinking water, reducing disease risk, and dampening storm waves and flooding)
- **Cultural services**: nonmaterial benefits (cultural identity, spiritual connections, aesthetics, recreation, and ecotourism opportunities)

Many of these services are essential for short-term human survival, and they all enhance people's quality of life.

As people seek to meet their needs and desires, they do things that affect ecosystems. These actions range from exploitation and impacts, such as pollution, which reduce the capacity of ecosystems to provide services to society, to stewardship, which sustains or enhances these capacities. If people degrade nature to meet their short-term desires, ecosystem services decline. This erodes society's capacity to meet its needs. On the other hand, if human well-being is severely compromised, people have no choice but to meet their immediate needs by taking whatever they can get from nature. In either case, both nature and people suffer. How do the positive and negative effects of people on ecosystems balance out at the global scale?

In 2005, 1,300 ecologists from around the world completed a 5-year global assessment of the impacts of ecosystem changes on human well-being. The resulting Millennium Ecosystem Assessment (MEA) concluded that two-thirds of Earth's ecosystem services were being degraded or used unsustainably in the global aggregate. One-fifth of ecosystem services had not been systematically altered. Only three (13%) actively managed services (crops, livestock, and aquaculture) were increasing (Table 3.1).[5] A 2019 update of this synthesis shows that 75% of ecosystem services are now being used unsustainably.[6] These statistics clearly show that the services provided by most ecosystems are declining in their capacity to meet human needs.

Most of the declines in ecosystem services result from human impacts on ecosystems.[7] Pollinators that are essential for fruit production are declining because of insecticides and pollution (Chapter 2). Human introductions of invasive species have reduced the capacity of many grazing lands to support cattle. Drainage of wetlands for agriculture and development has reduced their capacity to remove pollutants from agricultural runoff and to buffer the impacts of heavy rains on downstream flooding.

By identifying the major causes of change in each ecosystem service (Table 3.1), the MEA provides clear guidance about ways to foster their recovery— simply reduce the pressures that cause their declines. This approach is generally a safe starting point for reducing environmental degradation because it

Table 3.1 Global Status of Ecosystem Services

Ecosystem Service	Status	Causes of Change
Provisioning services		
Food crops	Increasing	Substantial production increase
Livestock	Increasing	Substantial production increase
Aquaculture	Increasing	Substantial production increase
Capture fisheries	Decreasing	Declining production due to overharvest
Wild foods	Decreasing	Declining production due to overharvest and habitat loss
Timber	Variable	Forest loss in some regions, growth in others
Cotton, hemp, silk	Variable	Declining production of some fibers, growth in others
Wood fuel	Decreasing	Declining production due to overharvest and forest loss
Genetic resources	Decreasing	Loss through extinction of species and crop varieties
Biochemicals, pharmaceuticals	Decreasing	Loss through extinction and overharvest
Fresh water	Decreasing	Unsustainable use for irrigation and industry
Regulating services		
Air-quality regulation	Decreasing	Declining capacity of the atmosphere to cleanse itself
Climate regulation		
Global	Decreasing	Less carbon capture by ecosystems due to forest loss
Regional, local	Decreasing	Preponderance of negative impacts
Water regulation	Variable	Depends on ecosystem change and location
Erosion regulation	Decreasing	Increased soil degradation from poor farming practices
Water purification	Decreasing	Water pollution and loss of natural wetlands
Disease regulation	Variable	Depends on ecosystem change
Pest regulation	Decreasing	Natural control degraded by pesticide use
Pollination	Decreasing	Global decline in pollinator abundance
Hazard regulation	Decreasing	Loss of natural buffers (wetlands, mangroves)
Cultural services		
Spiritual values	Decreasing	Rapid decline in sacred groves and species
Aesthetic values	Decreasing	Decline in quantity and quality of natural lands
Recreation, ecotourism	Variable	More areas accessible but many degraded

addresses the root causes of problems. The necessary steps are generally well known by resource and restoration managers—reducing rates of forest and wetland loss and restoring or replacing those forests and wetlands that were previously eliminated. Reducing the declines in ecosystem services is therefore not conceptually complex. The difficulties arise largely from **trade-offs**— when gains in some ecosystem services, such as agricultural production, come at the expense of losses of other services, such as climate regulation by forests.

Balancing Nature's Benefits

Land-clearing for agriculture and forest harvest is the largest cause of the global decline in ecosystem services.[8] This began thousands of years ago but has accelerated in the last 200 years—especially since World War II. We annually harvest 40% of Earth's productivity on land and divert half its fresh water—mostly to support agriculture—rates that cannot be sustained by natural processes.

How can society minimize the trade-off between harvesting ecosystem products to meet society's immediate needs and the other important services that nature provides? Reducing human demands for food and other natural resources is the least risky way to reduce pressure on other ecosystem services. As Pope Francis of the Catholic Church pointed out,[9] the most effective way to reduce human impacts on our planet is for people in developed nations to reduce their consumption of energy and other natural resources. In the United States and Europe, the average person consumes 32 times more energy and other natural resources than does someone in a developing nation.[10] That's more than our fair share. A decline in the rate at which human population increases would further reduce society's consumption of resources.

Reducing the resources consumed per individual is a complex challenge that I will address in Chapter 6. As a teaser, I will simply say that reducing individual consumption is feasible. If done thoughtfully, this would probably increase the well-being and happiness of most people in developed nations in both the short run and the long run—but that's another story (Chapter 6).

Now, on to the population issue. Throughout the world, especially in poor countries, people are migrating to cities in search of better lives. Over half of today's human population lives in cities.[11] This proportion is rapidly rising, especially in Asia and Africa. In addition, some countries, like China and India, have brought millions of people out of poverty into a middle class. In both circumstances, large families become an economic burden rather than an essential source of labor and old-age insurance. In addition, women are taking charge of their lives. More women choose to work. These demographic and economic changes create incentives for families to have fewer children. Moreover, in China, with its policy of one child per family, a generation of families became accustomed to the personal flexibility resulting from small families. Most Chinese families therefore continue to have few children, even though the one-child policy has been relaxed.

For these and other reasons, the rate of human population growth is declining, and population is expected to peak around the end of this century. This would bring an end to the exponential human population growth of recent centuries.[12] That's good news. But will this shift happen quickly enough to avert potential catastrophes resulting from having more people on our already crowded planet? By then, we will have 50% more people than today, and we already have 2–3 times more people than the planet has been able to sustainably support. What morally defensible actions can be taken now to reduce the population pressure on our planet?

The most rapid population growth rates are in sub-Saharan Africa, which is where most future population growth is expected to occur. Here, people often want many children, in part because children still provide valuable farm labor and security for aging parents. Parents also remember vividly the history— and, for many, the current reality—of high infant mortality as they plan their families. The relationships between men and women are deeply embedded in culture and therefore sensitive topics to incorporate into international development efforts. A less controversial approach has been to improve access to general education for girls. This has proven to be the most effective intervention in reducing high population growth rates in this region.[13] Girls then have opportunities to learn about health and to gain skills. This opens opportunities for them to make money for themselves, imagine a future they would choose, and provides a broader understanding of human rights and cultural contexts. This win–win intervention both enhances women's opportunities to accomplish personal goals and reduces population growth in a region where addition of more people would increase starvation risks.

In all countries, rich and poor alike, improved access to information and support for family planning reduces the likelihood of unwanted pregnancies. This access is especially important for young people who are just entering reproductive age so that they can make responsible decisions that will strongly shape the rest of their lives. Reproductive choices are complex because they are both highly personal and influenced by the religious, social, and political contexts in which people live. They also influence, especially in developed nations, the impact of society on the planet.[14] The least we can do is to empower people to make their own well-informed choices within a complex social environment.

The need for agricultural land can also be reduced by increasing the amount of food produced by existing farmlands. Both conventional crop-breeding programs and genetic engineering have greatly increased the productive potential of crops—about 5-fold for cereal grains.[15] Cereal yield in the United States increased about 3-fold from 1961 to 2014 despite a slight decline in land area under cereal cultivation. However, this high productivity required more water, fertilizers, and pesticides. Intensification of agriculture affects other ecosystems and creates trade-offs with ecosystem services. For example, pesticides increase health risks to farmworkers and consumers and reduce pollination by bees. Both pesticides and nutrients can contaminate drinking water and pollute downstream ecosystems. But it doesn't need to be this way.

Pamela Matson is an ecosystem ecologist and sustainability scientist who studies factors affecting the cycling of nitrogen and other nutrients in ecosystems. She realized that agriculture and especially the high rates of nitrogen fertilization were harming environmental and human health. She therefore looked for ways to produce food without these unintended consequences. With colleagues from several disciplines, she tackled this issue in the Yaqui Valley, Mexico—home of the green revolution where intensive agriculture was born.[16] They found that farmers added much more fertilizer than necessary to support high yields and that overfertilization resulted in nitrogen losses that affected not only land, air, and water environments but also farmers' pocketbooks. However, farmers' decisions were not the root of the problem. Their fertilizer use was based on "advice" from farmers' unions. These unions provided credit—loans needed by farmers to buy seed, fertilizer, and other inputs—and their advice included a requirement to use large amounts of fertilizer as a relatively cheap (in dollars) way to ensure maximum crop yields. By working with the credit unions as well as farmers, Pam and her colleagues found ways to reduce fertilizer inputs, save farmers money, and reduce nitrogen pollution to the Gulf of California and elsewhere.

Pam's husband, Peter Vitousek, is another ecosystem ecologist who used ecological understanding to reduce pollution from intensive agriculture. Peter worked with colleagues in China, where government has substantial influence over farming practices. His Chinese colleagues examined a variety of farming practices to test their match of nutrient additions with crop nutrient demand.[17] They identified practices that could achieve yields even greater than in intensive agriculture but with much less fertilizer and pollution (Chapter 2). Greater attention to local variation in economics, environment, and culture can identify incentives and practices that could globally reduce the pollution impacts of agriculture.

Dietary changes have created an agricultural trend that threatens to cancel out the land-sparing benefits of intensive agriculture. As people in developing nations like India and China become more affluent, they consume more meat. Animals—especially warm-blooded animals like cattle—require about 7 times more feed than they produce as meat for market.[18] Cattle require 10–15 times more land per gram of protein produced than do poultry, which require 2–8 times more land than do grain and vegetables. Currently, about 80% of global agricultural land is used to raise livestock.[19] About half of this is unimproved grasslands and rangelands with limited agricultural potential.[20] The remainder is pastures for grazing and croplands that produce animal feedstocks. It takes a lot of corn and other grains to feed animals—46% of harvested crops![21] A shift to less meat-intensive diets would reduce the land requirement to feed people.

A final solution pathway to reducing land needed for agriculture is to increase the likelihood that harvested food actually reaches and feeds the people who need it. Agriculture now produces enough food to feed all the people on the planet, but much of this food is wasted. In developing nations, most food waste occurs because of inadequate food storage and delivery systems.[22] In addition, most people have strong preferences for certain foods, so the food that's available to them may not be culturally acceptable.

In developed nations, however, food waste results primarily from food that is thrown away. About 40% of food produced in the United States and Canada is thrown away by individual consumers. Most of this ends up in landfills, where it decomposes and releases **greenhouse gases** (gases like carbon dioxide [CO_2] and methane that trap heat in the atmosphere and warm the climate—Chapter 4). For all these reasons, about a third of globally harvested food is wasted rather than being eaten.

Food distribution is also shaped by markets and government trade policies. People and nations typically sell food at prices determined by local and global markets—prices that poor people and nations may not be able to afford. Trade policies complicate the economics of food distribution by adding subsidies and penalties that reflect other national goals. Energy laws requiring greater use of non-fossil-fuel energy, for example, increased the demand for corn and palm oil.[23] A third of US corn production is now used to make industrial products such as biofuels. The resulting price increase of corn benefits farmers but makes it less affordable to people in poor countries. Ironically, the price increase also stimulated legal and illegal land-clearing for oil palm plantations in the tropics and the return of soil conservation lands to corn production in the temperate zone. Both of these land-use changes released CO_2 to the atmosphere and offset much of the intended benefits of replacing fossil fuels with plant-based fuels.

Brazil successfully reduced its rate of deforestation through several efforts. A government anti-poverty initiative launched in 2002 sought to sustain the livelihoods of indigenous forest peoples through policies that reduced deforestation. Meanwhile, national and international nongovernmental organizations publicized the role of Brazilian agricultural exports (especially soybeans by Cargill and beef by McDonald's) that were economic incentives for deforestation.[24] As a result of this exposé and the reactions of North American buyers, Brazilian grain and beef exporters announced a voluntary moratorium on the purchase of soy and beef from recently deforested lands. Finally, Norway pledged up to $1 billion in compensation to Brazil for proven reductions in deforestation. These actions turned the incentive structure for deforestation on its head—incentives now favored protection of Brazil's tropical forests. As a result, the annual deforestation in Brazil declined 5.5-fold from 2004 to 2014. Unfortunately, changes in forest protection laws and 2018 promises by newly elected president Jair Bolsonaro to promote forest-clearing have led to increased deforestation in recent years, demonstrating the sensitivity of ecosystem protection to political changes.

In contrast to Brazil, the Indonesian government has done less to curb deforestation.[25] There is little transparency in either its statistics or policies related to forest changes. Much of Indonesian deforestation is illegal, often on government-"protected" carbon-rich wetlands. This deforestation supports the establishment of oil-palm plantations that account for 11% of the country's export earnings. China and India import much of this palm oil. Comparison of Brazil and Indonesia shows the importance of differences in both national policies and international trade in shaping patterns of deforestation.

Climate negotiations have created agreements—such as the United Nations program Reducing Emissions from Deforestation and Forest Degradation (REDD+)—that provide economic incentives for reducing deforestation. Large CO_2 emitters like the United States can meet part of their climate obligations by paying other countries to reduce their rates of deforestation.[26] One challenge with this approach is to ensure that the money reaches local landowners rather than paying rich landowners or the politicians who negotiate the deals. It is also important to curtail the illegal logging that often continues despite international agreements—hence the importance of Norwegian insistence of *proven* declines in deforestation as a condition of their compensation to Brazil.

Some people harvest food from ecosystems without substantially altering natural ecosystem structure and functioning. These include many cattle ranchers in savannas of Australia, the United States, and Argentina; nomadic herders in Africa and central Asia; small-scale artisanal fishers; and indigenous peoples of many nations. Grass-fed cattle that are raised on rangelands provide health and conservation benefits and an economic incentive to conserve rather than to develop these lands. People from many cultures and segments of society pick berries, collect mushrooms, or hunt and fish in natural landscapes for the nutritional, recreational, and cultural values of these products and the pure pleasure of immersing themselves in nature.

Clearly, natural and seminatural ecosystems are an important part of the world's food system, both on land and in the sea, especially for people who are not fully integrated into the market economy. People who rely significantly on food from natural ecosystems account for most of the world's linguistic and cultural diversity. Although native peoples constitute only a few percent of the world's population, they inhabit about 25% of Earth's land area—areas that account for most of the world's species diversity.[27] These are the people with the greatest diversity of understanding about how to interact with local nature, even though they may not be rich or powerful enough to compete strongly in the market economy.

Trade-offs between food production and other ecosystem services are just as profound in the ocean as on land. Aquaculture is the marine counterpart of terrestrial agriculture and has analogous impacts on ecosystem services. Shrimp farming in Southeast Asia expanded by clearing mangrove forests that would otherwise protect coastlines from storms.[28]

Just like cattle on land, fish consume more calories than they convert into their own body mass, so more fish or other food must be harvested (generally from the open ocean) to feed aquaculture than can be brought to market as aquaculture products.[29] Some of the unintended side effects of aquaculture are similar to those in terrestrial stockyards.[30] Both systems pollute surrounding waters because they produce more wastes than can be locally recycled. In addition, the antibiotics used to control diseases at high animal densities become a health risk both to people who eat the meat and to other organisms that live in the waste stream of these operations. Excessive use of antibiotics promotes the evolution of antibiotic-resistant bacterial strains that are extremely difficult to control. Some aquaculture can be supported by plant-based feeds or algae or can recycle waste streams into fish or other products. Food production by aquaculture could therefore potentially increase food production without polluting or putting pressure on wild fish stocks. However, this is currently not the typical pattern.

Wild-capture fisheries account for about half of the food harvested from the oceans. These fisheries provide an important protein source for people and reduce pressure for land-based agriculture. Total marine fish harvest has declined since 1980 despite increased fishing pressure because of widespread overharvest.[31] The largest number of these fisheries are not actively managed and are fully or overexploited. However, this outcome isn't inevitable. Large global fisheries, such as those for pollock and anchoveta, tend to be managed scientifically; and their stocks are increasing.[32] These patterns suggest that full use of good management practices known today could make most fisheries sustainable and would increase global fish harvest, while increasing fishery stocks.

Protection of Society by Ecosystems

People are connected to nature through more than food, fiber, and water. Ecosystems also mediate interactions among ecosystems by regulating the flow of water, nutrients, and pollutants from their origins to other places, as the city of Houston learned from its increased vulnerability to flooding (Chapter 2). Unlike the obvious economic benefits of materials that we harvest from ecosystems, the economic value of regulating services is less visible to society because we are used to getting these services without paying for them. As these services decline due to human actions, their value to society becomes more obvious.

A relatively new strategy for conserving or restoring regulating services has been to estimate their economic value—not because this is the only reason we care about them but because economic valuation gives ecosystems a seat at the table when hard-nosed economic decisions are made.[33] When these values are entered into a balance sheet, the value of ecosystem services to society becomes explicit.

Until recently, there was no marketplace where the people who most benefit from these regulating services could pay those who take actions to sustain these services. Explicit calculations of the economic value of ecosystems' capacity to regulate water quantity and quality can lead to better decisions—both ecologically and socially.

New York City historically depended on clean water from the Catskill watershed.[34] For the past century, the city reduced water pollution by helping watershed residents improve septic and sewage treatment facilities and purchasing development rights. The resulting habitat protection maintained recreational opportunities and property values in the affected watersheds. Nature protection was also less expensive than building a water filtration plant in the city.

This model of investing in watersheds for downstream water benefits is gaining traction. Over 30 Latin American cities have such programs in action or under development.[35] The InterAmerican Development Bank used this model in a water infrastructure loan to Brazil where 10% of the loan was committed to watershed investment because of the believed benefits for downstream water security.

Fort Collins, Colorado, is proud of its water and its beer. Devastating wildfires in 2012 hit close to home when the High Park Fire was followed the next year by a 100-year flood that inundated the town with ash and mud. Local microbreweries that depend on clean water to make good beer teamed up with The Nature Conservancy to organize a community-wide Octoberfest celebration to raise community awareness about the links between healthy mountain forests and the town's economy and quality of life. Community awareness morphed into the Peaks to People Water Fund.[36] Community donors to this fund pay mountain-land stewards to manage their lands in ways that reduce the fire risk to fire-prone forests or reduce sediment delivery to the river from recent fires. This fund connects a community that is proud of its water with private landowners who are proud that their lands can deliver this ecosystem service.

Regulating services are also important to coffee lovers. Natural ecosystems adjacent to coffee plantations in Costa Rica contribute more to coffee production as habitat for pollinators and pest-regulating insects than as additional land for growing coffee.[37] Once the economic values of less visible ecosystem services are known, payment schemes can be developed that compensate landowners for keeping lands in natural condition, as in the New York City and Costa Rican examples.

Regulating services are ecologically and societally important because they link decisions made in one place with people in another place—through climate, fire, or water connections. Economists generally agree that societal costs (**externalities**) should be included when costs and prices are calculated so that society isn't left paying the bill or suffering the consequences (Chapter 8). It's sad that wetlands—which accumulate water during floods and release it slowly after flood conditions have passed—are being drained most rapidly near cities. This is precisely where most people need the protection provided by these ecosystems.

Low awareness of the importance of regulating services also influences the behavior of farmers. Midwestern farmers usually apply more nutrients to their fields than crops absorb—just as in the Yaqui Valley—with the excess nutrients going downstream and creating dead zones where rivers meet the sea.[38] If there is no mechanism for a farmer to recognize and pay (or be paid) for the cost of the lost coastal fishery, he or she may decide that preventing pollution is not cost-effective.

Sustaining Cultural Links to Nature

Bill McDonald is a fifth-generation rancher in the Malpai Borderlands of southwest New Mexico and southeast Arizona.[39] He and neighboring ranchers were concerned that conservation groups were buying up ranches and excluding cattle in the name of nature protection. In other places, federal land managers suppressed wildfires on both public and private lands. Yet Bill knew that fires usually increased grass cover and enhanced habitat for both cattle and rare plant species. When managed wisely, both fire and cattle-grazing supported conservation and reduced fire risk. After decades of bitter feuds, Bill and a few others brought together ranchers, conservationists, and federal land managers of the region around their greatest shared concern—that ranch lands would give way to a development patchwork that would not meet the goals of any group. They formed the Malpai Borderlands Group to address the role and future of ranching in the region. Once they trusted one another enough to work and learn together, they found that their distinct perspectives and experience simultaneously improved opportunities for ranching, conservation, and land stewardship. The cultural and economic benefits of these ecosystems provided the foundation for solutions. Similar stories emerge wherever people have deep connections to their local land or sea and make the effort to develop trust and collaborate (Chapter 8).

Changing regional economies often challenge the economic viability of ranching. Range-fed cattle are expensive to produce in terms of labor and land, relative to beef from feedlots or imported from other countries. Despite the difficulties of ranching livelihoods, most ranches between the Rocky and Sierra Mountain ranges of the United States are still operated by the families who have done so for generations. Malpai ranchers joke that "Ranching is a genetic defect—something you inherit and can't get rid of, no matter how hard you try."[40] Ranching is more than just a livelihood. It is a way of life that is central to the culture and identity of ranchers and other residents of the region. The cultural benefits of these lands are just as valuable to this community as is the forage on which their cattle depend. As these regions increasingly attract retired people or telecommuters, ranch lands fetch high prices when sold for housing developments and hobby ranches. Rising property values tempt ranchers to sell their lands to developers or face higher taxes. Conservation easements can reduce the tax burden and incentives to sell, making it easier for ranchers to sustain cultural connections to their lands.

My wife's two uncles were ninth-generation farmers in western Massachusetts. The lowlands of their farm were planted to pasture for summer grazing and corn for silage that fed the dairy herd in winter. The rocky uplands were forests that supplied sap for maple syrup and logs for the family sawmill. Each uncle drew his identity from a different thread of family heritage. One focused on efficient management of the dairy farm and the other on managing the woodlot and making maple syrup, supplemented by an off-farm job in a nearby town. During much of their adult lives, the uncles argued bitterly about whether to manage the farm for agriculture or conservation. When they retired, they split the farm between them into fields and woods. Without talking to one another, the farming uncle sold the development rights to his agricultural land, committing it to farming in perpetuity rather than to housing developments. The woodsman uncle set up a conservation easement that committed his woods to forestry-based livelihoods and nature conservation. Although they couldn't agree how the land should be used, both chose to perpetuate a land ethic that had shaped their separate lives.

The uses of rural lands also affect society at large. The health and environmental benefits of grass-fed beef compared to cows that are fattened in stockyards are well known. In addition, croplands in central California have 2.5- to 3-fold higher emissions (in CO_2 equivalents) per land area than do rangelands and associated stockyards.[41] Conversion of cropland to urban and suburban uses causes a further *70-fold* increase in emissions! Clearly, conversion of farms and rangelands to other land uses has societal consequences that extend well beyond the culture and identity of farmers and ranchers.

Indigenous people also have a multigenerational view of their relationship to the land. AlexAnna Salmon and her sister Christina grew up in the remote Yupik village of Igiugig in southwest Alaska. Their father, a community leader, died in a plane crash during AlexAnna's senior year at Dartmouth College. An elder in her community wrote to her, "Please come back and help our village." AlexAnna decided to shelve her plans for graduate school and returned to Igiugig to continue the community work her father had begun.[42]

I met AlexAnna, Christina, and their friends through a research collaboration between the Igiugig Tribal Council and the University of Alaska Fairbanks (Chapter 7). The main goals of the Igiugig Tribal Council, as in many native communities, are to sustain their cultural integrity and ties to the land. They tell me that they have spent enough time in cities to know that, for them, the benefits of living on the land far exceed those of city life. The fish, berries, and other foods harvested and shared each year are much more than food—they are a chance to spend time on the land with family and friends.

Most villages in rural Alaska have a high unemployment rate, and half the households there are below the US poverty line. Life is economically tough because there aren't many jobs, and commercial goods, which must be flown in, are expensive—$6–$10 for a gallon of milk or gasoline. Therefore, all families depend on harvesting food from the land: salmon, moose, berries.

These people work as hard as anyone I know, but they do this in the context of a life on the land. During salmon runs, they and their kids and grandparents are on the river catching and cutting fish—enough for themselves and to share with neighbors who can't leave their jobs or are too old to fish. They wrote grant applications that supported a wind- and solar-powered greenhouse. Vegetables from the greenhouse and chickens raised on food wastes supply healthy local food for school lunches. Elders often join the kids for lunch, strengthening both nutritional and cultural ties across generations. Christina campaigned and was elected to the Borough Assembly. Here, she advocates for the rights of local people and the clean water and salmon on which they depend—rights that are threatened by plans for large-scale mining in the watershed. My favorite Facebook posts are from AlexAnna and Christina because of the fun times they show with their kids, friends, and community elders. They work hard, but they laugh a lot. They have figured out how to live well.

Most people have a richer life because of the cultural and recreational benefits they receive from ecosystems. Cities would be sterile places if they had no parks or street trees, where people can play and socialize (a cultural service). Even people who pay no attention to urban nature benefit from it. It cools the urban heat island, which reduces the risk of heatwave fatalities (a regulating service, Chapter 4). It also prolongs the life of the elderly—for example, reducing mortality of people over 80 by nearly 30% in Chinese cities.[43] Urban gardens, like those in Stockholm (Chapter 1), sustain the rural memories of people who moved to cities and want to share these memories with their grandchildren.

Children all over the world have empathy for the future of polar bears, even if they never expect to see one. Our connection to nature matters in ways that far surpass its immediate economic benefits.

What Can We Do?

There are many options for actions to foster more favorable interactions between people and nature:

- **Celebrate and protect the ecosystem services that shape our community's identity.** Identify, and discuss with others, ways in which these ecosystem services benefit the community.
- **Identify critical threats to ecosystems valued by our community.** Consider actions that would make valued ecosystems more resilient to changes that are happening.
- **Learn how global changes affect valued local ecosystems.** These interactions might include climate change (Chapter 4), pollution, and land-use change.
- **Take actions that benefit ecosystem services globally.** Consider changing our diet and energy use to habits that reduce agricultural land needs.

The next question is: *Given the impact of climate change on the well-being of people and nature, how can these changes be understood and their negative effects minimized?*

4

Tackling Climate Change

Rapid climate change affects both nature and society. This chapter describes actions that people can take to understand and minimize climate change.

Causes of Changes

Climate warming has now become—along with land-use change—one of the two leading causes of the ecosystem degradation that harms both nature and society.[1]

Alexander von Humboldt, in 1800, was the first person to suggest that human activities could alter Earth's climate.[2] He was a true Renaissance man. His goal was to bring together all knowledge of science and culture into a unified cosmos of understanding. As a young man, he studied many branches of knowledge with Europe's leading experts. He traveled throughout Europe to learn about the geographic patterns of the topics he studied. In 1799, he began a 5-year exploration of the Americas and surrounding seas. He documented the relationships among large-scale and local patterns of climate, Earth's gravity and magnetism, geology, vegetation, and culture. Among other discoveries, he learned that the geographic patterns of climate largely explained both global and altitudinal distributions of vegetation structure and composition—a forerunner of Hans Jenny's critical-factor framework for ecosystems. Archeological ruins suggested to Humboldt that past changes in climate had contributed to the rise and fall of civilizations. I was amazed to learn that Humboldt had figured out the major causes and consequences of climate change more than 200 years ago. The reasons for these changes are now quite well understood (Box 4.1).[3]

Box 4.1 **Basics of Climate Warming**

Although Earth's climate system is complex, the basic principles that explain climate warming are fairly easy to understand.[4] Here's how it works.

The sun, because of its intense heat, produces high-energy radiation, about half of which penetrates Earth's atmosphere and warms Earth's surface (Figure 4.1).[5] Earth, in turn, releases about the same amount of energy back to the atmosphere but does so as lower-energy radiation (heat) because of its cooler surface temperature. Most of the heat released by Earth is absorbed by the atmosphere's heat-trapping gases, especially by water vapor and CO_2. These greenhouse gases, as they are sometimes

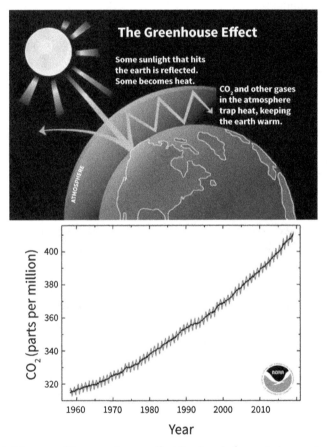

Figure 4.1 Diagram of the greenhouse effect (top) and of recent changes in CO_2 concentration in the atmosphere measured near the middle of the Pacific Ocean at Mauna Loa Observatory, Hawaii (bottom). Increases in atmospheric CO_2 concentration intensify the greenhouse effect by increasing the capacity of the atmosphere to absorb heat. Rising CO_2 concentration is the major cause of climate warming.

called, act as a blanket that moderates Earth's temperature enough for our planet to support life. That's a good thing. In addition, CO_2 is an essential input to photosynthesis, which supports ecosystem productivity. If there were no greenhouse gases or clouds, Earth's average surface temperature would be about –60°F. These basic principles have been understood since the 1800s, but we now see them play out in new ways.

During the 10,000 years before the Industrial Revolution, respiration released CO_2 from ecosystems to the atmosphere at about the same rate that plants removed CO_2 from the atmosphere to support ecosystem productivity. As a result of this balance between inputs and outputs, the atmosphere's CO_2 concentration remained fairly constant, stabilizing Earth's temperature (Figure 1.1).

Burning of fossil fuels now adds CO_2 to the atmosphere much faster than it is removed by natural processes, causing CO_2 to accumulate. We know that the new CO_2 in the atmosphere comes from fossil fuels because it has the same telltale ratio of forms of carbon (**isotopes**) as do fossil fuels. This isotope ratio of fossil fuels and of new atmospheric CO_2 differs from that of the CO_2 that is naturally released by ecosystems.

As the CO_2 concentration of the atmosphere rises due to burning of fossil fuels, Earth's atmospheric blanket has become more effective at retaining heat, causing the climate to warm.[6] Other heat-trapping gases, such as methane and nitrous oxide from intensive agriculture, also contribute to climate warming but accumulate less in the atmosphere than does CO_2 because they are broken down relatively quickly. These human-produced greenhouse gases absorb Earth's heat at different wavelengths than does water vapor, the other important greenhouse gas. As the human-produced greenhouse gases heat the atmosphere, the air can hold more water vapor, so more water evaporates, magnifying the overall rate of climate warming. Considering all of these changes together, CO_2 accounts for about two-thirds of the energy imbalance that causes Earth's climate to warm.[7]

Some people (18% of Americans) are doubtful or dismissive that the well-documented buildup of atmospheric carbon dioxide (CO_2) from burning of fossil fuels explains climate warming (Figure 4.2).[8] However, no scientist has offered an alternative explanation for the temporal patterns of warming that have occurred over the past century.[9] Other people (60% of Americans) are more deeply concerned about climate change.[10] Regardless of people's opinions about climate change, actions intended to reduce climate warming would have many benefits for society—as shown on the screen behind the speaker in Figure 4.2.

Figure 4.2 Some people believe that climate change is a hoax perpetrated by scientists, despite the many other benefits that would result from actions to reduce the rate of climate change. Cartoon by Joel Petit.

Climate is warming faster now than in the past.[11] About two-thirds of the warming since 1900 (1.0°F) has occurred since 1975, and 18 of the 19 warmest years in the instrumental record (since the late 1800s) have occurred since 2000.[12] Climate models project that global temperature will rise an additional 2.5–8°F by 2100. This wide range in possible future warming reflects uncertainty about whether and how quickly society will reduce its fossil-fuel emissions. In the absence of strong actions to reduce emissions, the world will get *much* warmer, and parts of it will become uninhabitable. The US military views climate change as one of today's greatest risks to American and global security because of its impacts on human safety, food security, political instability, and migration.[13]

Given the rapid recent and projected warming, people born this century will probably experience more climate warming during their lifetimes than has been seen by all of humanity since the beginning of major human civilizations (Chapter 10). Earth's climate is moving out of the range of the temperature stability that has nourished the development of agriculture, civilizations, cultures, and religions over thousands of years (Figure 4.3).

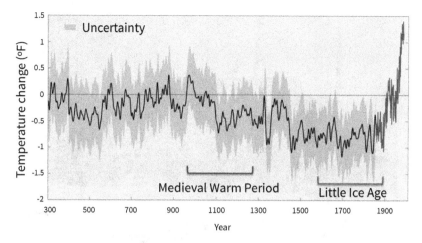

Figure 4.3 Changes in Northern Hemisphere air temperature over the last 1,700 years (relative to the average temperature 1961–1990). The temperature record from the years 300 to 1860 comes from proxy records such as those recorded in glacier ice. The record since 1860 comes from direct instrumental measurements.

Impacts of Changes

Climate change doesn't grab everyone's attention the way that dust storms of the 1930s did for Dust Bowl farmers. After all, the greatest changes in the climate system are invisible (CO_2, a transparent gas), distant (diffuse across the planet), and long-term—precisely the features that seem least urgent as people pursue their daily lives.

However, to Stanley Tom, who lives in the coastal village of Newtok, Alaska, climate change is *the* most immediate and urgent issue he faces every morning. Rapid warming in Alaska is thawing the permafrost that stabilizes the ground beneath Stanley's village and melting the sea ice that used to protect his village from waves during autumn storms (Figure 4.4).[14] The 200 feet of tundra that used to separate Newtok from the sea has eroded away. It may be only a matter of months before the school, which serves as the village evacuation center during storms, will be lost to coastal erosion.

Similarly, Tomás Regalado, former mayor of Miami, Florida, used to think of climate change as a distant future possibility until storm surges from Hurricane Irma in 2017 flooded cars and stranded residents of his city.[15] The majority of Miami's voters supported his 2017 bond, half of which was to protect Miami from rising sea level and climate change. Parts of Miami now flood almost monthly—a frequent reminder of the urgency of protecting the city against sea-level rise.

Figure 4.4 Climate-driven changes that threaten Alaskan villages include ponding of water from permafrost thaw and ground subsidence in Newtok (top) and damage to homes from fall storms in Shishmaref (bottom).

Warming causes sea level to rise in two ways.[16] First, warming causes water to expand and occupy more space, raising the height of the world's oceans. Second, warming causes land-based glaciers and ice sheets to melt, adding more water to the ocean.

Climate warming also heats the ocean surface, which then transfers more energy to tropical storms. Most climate models indicate that this would increase the frequency of intense hurricanes in a warmer world. This pattern is consistent with the international trend of increased frequency of intense hurricanes.[17] However, the satellite record, which provides the best record of these events, is too short to assess the significance of hurricane trends in any particular place.

The combination of higher sea level and more frequent intense storms increases the frequency of large storm surges in coastal areas. About 1%–2% of Americans and of global citizens live within 3 feet of mean high tide, making them particularly vulnerable to climate warming.[18] Global sea level has risen 7–8 inches since 1900 and is expected to rise by an additional 1–4 feet (possibly as much as 8 feet) by 2100,[19] making many coastal cities and other low-lying areas vulnerable to storm surges. Many people in these flood-vulnerable areas move to safer places after storms, as happened in New Orleans after Hurricane Katrina. Those who stay remain vulnerable to the next big storm.

Large floods in wet climates are also increasing under climate change.[20] Warm air evaporates more water and delivers it back to Earth as rain. Insurance companies are particularly strong advocates of climate research. Their own statistics show that climate-related property losses increased 5-fold between 1980 and 2016 as a result of both increases in extreme weather and building of new infrastructure in climate-vulnerable places.[21] These companies want to know where and how these risks will change in the future. As they raise their insurance rates to cover increasing climate-related losses, people with average or low incomes may no longer be able to afford home insurance.[22] Areas that used to flood once a century now flood more often—perhaps once every 10–50 years. Unfortunately, developers who look for inexpensive land often develop flood-vulnerable areas, as occurred in Houston (Chapter 2). However, changes in zoning regulations and insurance rates could discourage people from building in zones of high flood risk.

Choosing to ignore climate change is a costly proposition, and communities and their citizens are the big losers. The state of North Carolina passed legislation that prohibits the use of climate-change projections in development decisions, thus shifting the flood risks to individual homeowners and businesses.[23] Sea level in North Carolina is rising twice as fast as the global average because of a combination of rising seas and sinking coastal lands.[24] The resulting risk to property in just four counties is expected to be $6.9 billion over the next 75 years. Beaches in southern North Carolina are expected to lose $233 million per year in recreational income by 2080.

At a more personal level, one of my favorite childhood memories was bodysurfing on Carolina's broad beaches. Here, my father taught me how to catch and become one with waves that were much taller than me. The lazy waves would carry me shoreward as long as I could hold my breath and set me gently on the sand. After resting a moment to see if the water would carry me just a little further, I would jump up and follow the water back to sea in search of the next good wave.

By 2030 recreational beaches in southern North Carolina will have lost about half of their width to erosion associated with rising sea level, and by 2080 most of today's recreational beaches will have no beach at all.[25] Future beaches will develop where today's neighborhoods, forests, and other ecosystems will erode away.

August 2003 was Europe's hottest August on record—temperatures stayed around 99°F for a week in parts of France.[26] Heat records were again set in 2019. Cities are even warmer than the surrounding countryside because dark-colored roofs and pavement absorb more energy than vegetation, and cities have less vegetation to cool the air. The 2003 heatwave caused 35,000–70,000 more deaths in Europe than would normally have occurred. Heatwaves are not a rare phenomenon. Thirty percent of the world's population is exposed to potentially deadly heat for 20 or more days per year, on average.[27] Europe's hottest summers of the last 500 years have all come in the last 17 years.[28] If the climate continues to warm, extreme heatwaves will become more common, increasing health risks—especially for poor urban residents without air conditioning.

Warming also increases the risk of drought, which hurts everyone. In early 2018, Capetown, South Africa, was within 2 months of "Day Zero," when its water supply would run dry and the city would need to turn off most of its taps.[29] Although modest rains averted the immediate crisis, strict water rationing is a daily reminder of the city's vulnerability to drought. Even cities in wet climates, like Vancouver, Canada, face water rationing when the warmer climate melts snow packs that used to provide much of their summer water.

Climate change can't be blamed for any particular drought or flood. All we know is that extreme events—droughts in dry climates and floods in wet climates—are happening more often. This change is consistent with what would be expected with the climate warming that has occurred. These extreme events are the killers—not the modest changes in average temperatures that make these extremes more likely.

Warming also increases the risk of forest fires. In dry years, when most fires occur, wildfires are too extensive and intense to be fully controlled, even though the US Forest Service spends half or more of its budget on fire suppression. The most effective way to protect life and property from wildfire is to concentrate homes at a safe distance from flammable vegetation.[30] Widely dispersed homes in remote flammable places are dangerous and expensive to protect. Currently, most fire-suppression costs reflect protection of homes that were built in unsafe places rather than managing the changing ecology of fire. This dilemma could be avoided by changing zoning regulations and insurance rates to discourage people from building in unsafe areas—or, even better, reducing the rate of climate warming.

Actions to Reduce Climate Warming

Knowing that fossil-fuel emissions are the largest cause of recent climate warming is helpful. It shows clearly that the most effective way to reduce the rate of future warming is to reduce the emissions of CO_2 and other heat-trapping gases. This can be accomplished by reducing energy use and by drawing a larger proportion of our energy from renewable sources such as wind, solar, and hydropower (Chapter 6).

In addition to reducing emissions, CO_2 is directly removed from the atmosphere by vegetation. Forests, which store carbon in both wood and soil, and grasslands and wetlands, which store lots of carbon in soil, are particularly effective at removing CO_2 from the atmosphere. Protecting and restoring these and other habitats has the potential to achieve up to a third of the reductions in greenhouse gas emissions needed to meet recent (2015) climate targets and can be justified on many grounds in addition to their favorable climate impacts.[31]

Renewable-energy technology is developed enough to meet most of today's energy demands, if it were deployed more extensively. If regulations set a price on carbon emissions (Chapter 10), industry would have strong incentives for innovations that would make these technologies even more cost-effective and efficient. For example, coal, oil, and natural-gas companies might pursue innovations that scrub CO_2 from power-plant stacks to be pumped into deep geological reservoirs. Wind-energy and solar-energy companies might invest in battery or grid technologies for energy storage during times of low wind speed or solar radiation. Hydropower programs might focus on energy production that does not impede fish migration. Nuclear programs might focus on safety and reutilization of nuclear wastes. Government subsidies for all of these and other technological advances would speed innovation and deployment and reduce carbon emissions. There are thus both ecological and technological strategies to reduce the CO_2 concentration of the atmosphere. Both are urgently needed if we hope to prevent catastrophic climate change.

Other geoengineering approaches have been suggested as ways to reduce the rate of climate warming. However, these approaches are untested and potentially even more dangerous than current warming trends. Some people have suggested injecting tiny particles (sulfate aerosols) high into the atmosphere to reflect some of the sun's energy and therefore to shade and cool the Earth. Although sulfate aerosols would probably have this cooling effect, they would also reduce the sunlight needed to support plant production—especially of productive ecosystems such as agricultural crops—at a time when food needs will be even greater than they are today. Injected aerosols would also modify the climate system in different ways than does CO_2.[32] They might shift Asian monsoons in ways that create new patterns of drought and flooding. Sulfate injection would do nothing to relieve non-climatic effects of high CO_2, such as acidification of the ocean, which disrupts ocean food webs. Finally, any cooling effect of injecting particles into the stratosphere would be relatively short-lived and would commit society to continuing this intervention forever. Failing to do so for even a year would rapidly shift the climate to whatever superwarmed state was consistent with the CO_2 concentration at that future time.

The safest way to reverse climate warming is to reverse its current causes by both reducing CO_2 emissions and increasing CO_2 removal by vegetation. This is consistent with the general principal of reducing those human pressures that are responsible for current environmental and ecological degradation.

Climate-friendly strategies often have other locally valuable benefits (Figure 4.2), such as habitat restoration for threatened tropical-forest species and reduced severity of heatwaves in cities. None of these solution categories is conceptually complex. They just require the political will to tackle the climate problem (Chapters 6 and 9).

Globally significant climate action is not out of reach. There is an impressive array of groups already taking climate action in their communities. The Global Covenant of Mayors for Climate Action is a group of more than 10,000 communities and cities in 135 countries representing 11% of the world's population.[33] Each of these communities developed a climate action plan—starting with a few individuals who decided to address the climate issue in their community (Chapter 8). In addition, individual citizens can engage in climate action through lifestyles that use less fossil fuels and that include healthy diets with less meat (Chapter 6) and by supporting politicians who advocate for climate action (Chapter 9).

Climate scientist Katharine Hayhoe points out that the most important thing a person can do is to talk to other people about why climate action matters to them and what we can do to fix it. These reasons might include concern for the future of our children and for poor people who are particularly vulnerable to climate change, our spiritual commitment to care for

Creation, concern about the security of our nation and the world, or the economic opportunities needed by our community and nation (Chapter 7).[34] Renewable-energy industries now provide more new jobs than do fossil-fuel industries. Any of these reasons provides motivation to understand and take actions that reduce the rate of climate change. Reducing the rate of climate change is the most urgent challenge of our time.

Arctic Inuit people note that "the Earth is faster now."[35] Scientists who have studied the major causes and consequences of global changes come to the same conclusion. Society is in the midst of a great acceleration of changes that began after World War II—changes that are altering, ever more rapidly, Earth's characteristics and their consequences for society. The sooner society acts to reduce rates of climate change, the greater the likelihood of achieving desirable outcomes and reducing the risks of runaway climate catastrophes (Chapter 10).

What Can We Do?

People can reduce their impacts on future climate and their sensitivity to climate change in many ways:

- **Understand and minimize our personal impacts on the climate system**. Understand which of our actions most strongly affect climate, and consider how to reduce these impacts (Chapter 6). For example, explore ways to use less fossil fuels.
- **Talk with others about why climate change matters to society**. Explain why climate change concerns us personally and what we can do together.
- **Encourage elected representatives to pursue climate-friendly policies**. Actions could include voting for climate-friendly officials, writing to them to encourage climate-friendly actions, thanking them for constructive actions they have taken, and protesting government actions that increase rather than reduce rates of climate change (Chapter 9).
- **Work with others to reduce the climate vulnerability of our community**. What are the greatest climate-related risks to our community, and what actions might reduce this vulnerability?

The next question: *Given that people and nature depend on one another, how can society best meet its needs?*

5

Choosing to Live Well

All people have many similar needs to ensure a happy and satisfying life. This chapter describes how these needs can be sustained today and for future generations.

What Do People Need?

Abraham Maslow, the son of uneducated Jewish immigrants from Russia, grew up in Brooklyn, New York. He studied psychology during the Great Depression, as the events leading up to World War II and the Holocaust unfolded. And yet these devastating events during his youth did not dominate his perspective as a professional psychologist.[1]

Rather than following Sigmund Freud's preoccupation with psychological barriers to a satisfying life, Maslow sought to understand what allows some people to flourish. Maslow thought about the astounding accomplishments of Albert Einstein, Eleanor Roosevelt, and his own mentors.

Maslow described people's capacity to achieve their potential in terms of a set of universal human needs. He realized that some human needs, like food and water, are essential for life. He believed that other needs could be arranged in a hierarchy of importance. As people met their more basic needs, he thought they would be better able to address higher-level needs. This hierarchy included food and water; then feelings of safety, security, and health; then love and belonging through social relationships with family and community; then self-esteem and respect from others; and finally a person's capacity to realize her or his own potential and to address broader spiritual or societal goals.

Research during the years since Maslow proposed his theory has shown that actions by people throughout the world are indeed motivated by needs like the ones he identified. However, there is no universal hierarchy among these needs.[2] Some of the world's poorest people love their families passionately, are incredibly creative, seek opportunities to improve their children's lives, provide amazing leadership, and are motivated by spiritual values, despite difficulties in meeting their basic necessities. In addition, attitudes about needs differ among cultures, between individualistic and collectivist societies, among age groups, and in response to the social and political contexts of the times.[3] Young children often prioritize physical needs, older children love, adolescents esteem, young adults opportunities to achieve their potential, and older adults security.

Perhaps Maslow's greatest contribution was to point out the wide range of needs that motivate people's actions. Failure to meet any of these needs reduces a person's life satisfaction and capacity to contribute to her or his own goals and to those of society.

Reducing Poverty

Maslow's list of fundamental human needs surprised me when I first read it. It doesn't include material possessions or money, which are often used to distinguish "needy people" from others. What material possessions does a person really need?

La'ona DeWilde grew up in a cabin with her family 60 miles upriver from Huslia, Alaska, the village where I learned from Catherine Attla about her respect for the land. The DeWildes had little income and lived largely off what they fished, shot, trapped, gathered, and sewed. As a child, La'ona learned to snare rabbits with a piece of wire. She and her family joked that they lived off "the fat of the land"—the same land where, for 6,000 years, Athabascan Indians experienced cycles of sufficiency and starvation, depending on weather and shifting animal abundances.

Ben Stevens grew up 200 miles east of Huslia on the Yukon River on the same type of land. In winter his family hunted and trapped, and in summer they lived in a fish camp on the Yukon River, where they caught and dried fish for winter. In fish camp, Ben's grandfather told him that they were the richest people in the world. They had what they needed for a satisfying life.

La'ona's and Ben's childhood lifestyles are not what most people would choose. However, experiences of families like La'ona's and Ben's show that people's basic needs can't be measured by income alone. Well-being (or quality of life) also depends on people's expectations and the things they value.

Nonetheless, money provides a convenient metric to add up all of people's diverse needs. It's easier to add up dollars that someone pays for goods and services than to keep track of each separate need. For this reason, economists use money as a way to compare the poverty or well-being of different nations, segments of society, or periods of time. The problem is that people often forget that money is only a convenient measuring stick. Instead, money takes on a life of its own and seems inherently important to some people (Chapter 6).

Gross domestic product (GDP) is the measure of economic well-being commonly used to describe nations. It's the monetary value of all the goods that a country produces.[4] It is roughly equivalent to the total income earned or the value of the goods consumed.

The World Bank often uses GDP or income as a measure of a nation's poverty or well-being. The bank defines the poverty threshold in developing nations as the money needed by an adult to buy enough food to survive—$700 a year, $1.90 a day (in 2015), hardly a luxurious indicator of well-being.[5] Because of this emphasis on money as a convenient measure of well-being, international development programs put substantial effort into raising income and GDP as they try to bring nations out of poverty. Raising incomes of poor people is indeed important because it enables them to buy food and meet many other basic needs.

Income also matters to people and businesses that are not poor. There is therefore strong political pressure for policies that stimulate GDP—in other words increase the country's income through production of more goods and services. People would like to see their economy grow forever! However, this creates a problem. The associated increase in resource extraction needed to support greater production and consumption also degrades ecosystems and the services they provide to society.[6] Poor people are particularly vulnerable to losses of these ecosystem services (Chapter 3). Recognizing this dilemma, Nicholas Sarkozy, then-president of France, commissioned a report to recommend indicators that might improve international efforts to measure social progress and meet the needs of poor people. The commission concluded, in 2009, that achieving sustainable development requires direct measures of people's well-being rather than the current focus on income and other measures of production and consumption.[7]

Since the release of this report, there has been an explosion of research on ways to conceptualize and directly measure well-being.[8] The UK research program on Ecosystem Services and Poverty Alleviation summarized much of this research and concluded that well-being is more than an *outcome* such as a person's income.[9] It is a set of *processes* that influences what people can do with their lives.[10] These processes include

- sustaining the environmental and material conditions of a region so that people can thrive
- sustaining social relationships with family, friends, and community
- empowering individuals to meet their own goals

Historically, most international efforts to alleviate poverty have focused on providing the material conditions that poor people need but often fail to get— such as food, water, sanitation, disease prevention, electricity, infrastructure like roads, as well as the environmental conditions that support many of these needs.

People's social relations with family, friends, and community form a second dimension of well-being. Key social relationships depend on people's identity, what they value doing, and their freedom and security to pursue these relationships.[11]

The third dimension of well-being is more subjective—how individuals feel about their lives. Poor people who are asked what determines their well-being often emphasize their need to participate in society and to lead culturally meaningful lives. Even in dire material circumstances, people pursue these goals for their own well-being and that of their families.[12] They do the best they can with what they have, and this matters to them.

International aid efforts emphasize the first dimension of well-being— material and environmental conditions (Box 5.1).[13] This *outcome* improves people's immediate living conditions—greater income, less hunger, and improved health. Gross national happiness and genuine progress indicator (Chapter 10) are indices that incorporate additional dimensions of well-being. These indices identify actions and outcomes that would enable international aid programs to be more effective at meeting the needs of the poor people whom they are intended to serve.

Sustaining Foundations of Well-Being

International aid has an impressive track record of improving material conditions to support well-being (Box 5.1). The larger challenge, however, is to ensure that society can sustain these outcomes over the long run. I used to think that the well-being of citizens was the job of government or a product of a vibrant economy. Now I'm less convinced that reliance on government or the economy is sufficient.

Both government and the economy tend to focus on short-term rather than long-term goals. Elected officials must "deliver" outcomes to their constituents within a 2- to 4-year election cycle, if they want to be re-elected. Corporations must respond to daily stock market prices and justify their decisions to shareholders in quarterly reports. These short time frames motivate a focus on short-term outcomes rather than society's long-term needs

Box 5.1 **Progress Toward Sustainable Development**

In 2000, the United Nations set forth an ambitious agenda of eight Millennium Development Goals to improve the lot of humanity by meeting the needs of poor people. Here is a rough summary, as of 2015, of humanity's progress toward meeting these goals for all the world's people.[14]

Poverty and Hunger

1. **To eradicate extreme poverty and hunger**. The proportion of people living in extreme poverty declined from 47% to 14% between 1990 and 2015, largely through a 3-fold increase in the number of people in the middle class—those living on more than $4 a day—mostly in China and India. Twelve percent of the global population (836 million people) still live in extreme poverty. Among developing nations, the proportion of undernourished people dropped by almost half since 1990, but this decline was offset by population increases, so the *number* of hungry people did not change. The gap in wealth between rich and poor people has increased both within and among countries.

Environmental and Human Health

1. **To reduce child mortality**. The global under-5 mortality rate dropped by more than half between 1990 and 2015. However, in 2015, 5.8 million children under 5 died, mostly from preventable causes. Thus, child survival remains a significant global problem. Children of mothers with secondary or higher education are almost 3 times more likely to survive than children of mothers with no education.
2. **To improve maternal health**. Since 1990, maternal mortality fell by 45% worldwide, but even today only half of the world's pregnant women receive the recommended prenatal care.
3. **To combat HIV/AIDS, malaria and other diseases**. New human immunodeficiency virus (HIV) infections fell by about 40% between 2000 and 2013, but about 35 million people still live with HIV. Between 2000 and 2015, the expansion of malaria interventions led to a 58% decline in malaria mortality globally. Leprosy cases declined 97% between 1985 and 2017.
4. **To ensure environmental sustainability**. The net rate of forest loss dropped from 8.3 million hectares annually in the 1990s to 5.2 million hectares annually between 2000 and 2010. Over the same time period, global emissions of carbon dioxide increased by over 50%. Worldwide, 2.1 billion people gained access to improved sanitation. The number of people living in urban slums increased 28% from 1990 to 2015. The path toward environmental sustainability and human well-being is thus a mixed bag.

5. **To develop a global partnership for development.** Official development assistance from developed nations increased 66% in real terms between 2000 and 2014, although aid has plateaued since 2010. The debt burden of developing countries declined 4-fold between 2000 and 2013.

Education

1. **To achieve universal primary education.** Global enrollment in primary education increased from 83% in 2000 to 91% in 2015, still leaving 57 million children of primary-school age out of school. In developing regions, children in the poorest households are 4 times more likely to be out of school than those in the richest households.

Empowering People to Act

1. **To promote gender equality and empower women.** Most countries, including developing nations as a whole, have eliminated gender disparity in primary, secondary, and tertiary education—an amazing accomplishment. About half of working-age women participate in the global labor force, compared to three-quarters of working-age men, indicating a persistent gender disparity through life.

and well-being. It therefore falls largely to citizens and the nongovernmental organizations they support to advocate for longer-term needs.

Hans Jenny identified slowly changing factors, such as climate, soils, diversity, and disturbance, that sustain ecosystem structure and dynamics over the long run (Chapter 2). Analogously, we can identify foundations for human well-being (Figure 5.1). These *foundations* support the well-being *processes* identified by social scientists and give rise to the well-being *outcomes* articulated by Maslow. This framework of foundations, processes, and outcomes gives guidance of ways to sustain well-being over the long run: sustain or enhance the foundations of well-being, facilitate the processes that deliver this well-being, and monitor the resulting outcomes to assess progress.

Foundations of human well-being that are clearly linked to well-being processes and outcomes include the following (Figure 5.1):

- values that shape people's goals
- education
- cultural and social rules that sustain social relationships
- environmental and human health
- technology and economy

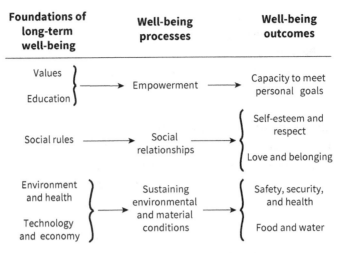

Figure 5.1 The primary relationships among long-term foundations, processes, and immediate outcomes that determine human well-being. There are also important interactions among these foundations and processes that are not shown but that are crucial in delivering desired outcomes.

Values and Goals

Values shape individuals' behavior that society considers appropriate. These include the tenets of religious faith and the values that shape people's relationships with one another and with nature—that is, the right things to do. Deeply embedded religious and moral values often bridge multiple generations and therefore provide a foundation for the long-term well-being of society. Values are complex products of human instincts both to compete and to care for others and for the environment. Within and among nations people differ in their attitudes about rugged individualism versus empathy and care for others. People also differ in whether they value group loyalty more than deeper moral codes. Values don't necessarily guarantee or undermine the foundations of well-being, but they do provide a slowly changing undercurrent that tends to sustain past motivations, attitudes, and practices.

Most societies value empathy. It unites the heart with the mind and is a core teaching of religions and moral philosophies.[15] Katrina Brown, a social geographer who studies poverty reduction in developing nations, first pointed out to me the importance of empathy.[16] Empathy extends care for others to people who may not be part of a person's network of friends and kin. Empathy leads a person to understand the emotions of others and indirectly to experience their emotional responses to conditions they cannot control. In this way, empathy fosters understanding and a willingness to care for the well-being of society's vulnerable people. In the absence of empathy, some people expect vulnerable people to solve their own problems and take care of themselves. Because empathy builds the capacity to act on behalf of others, it builds social consciousness and tends to reduce prejudice among groups. In general, empathy fosters attitudes and actions that benefit a broad spectrum of society.[17]

People tend to assimilate the attitudes and values of the people they trust and admire—parents, role models, and groups that are important to their identity (Chapters 6 and 7). In this same way, each of us influences the attitudes and values of others—our kids, friends, and many others we may not be aware of. Similarly, groups that focus on religion, environment, sports, and many other issues play important roles in promoting and sustaining values among their members.

Adults often complain that children no longer adhere to the values that they (the parents) learned as children. At the same time, children wonder why their parents are old fogeys who won't adjust to the modern world. As a child, I didn't understand my parents' frugal spending habits. Why didn't we have a television when all of our neighbors did? Why did we walk to school, when it would have been easier to go by car? We never talked about these things. Only later did I learn that all of my grandparents had lost their jobs or property during the Great Depression. The resulting collapse in family income instilled in my parents a deep distrust of the national economy and shaped their choices throughout life. Their reluctance to spend money on unnecessary things is a habit I assimilated, although my choices are motivated more by environment and future generations than by economics. Regardless of motivation, these values, attitudes, and habits tended to persist.

Migration from rural areas to cities often exposes people to radically different attitudes, habits, and values. City people engage in activities and gauge their neighbors' success quite differently than do rural people. Newcomers may therefore seek to emulate the spending habits and lifestyles of their new urban neighbors without fully understanding or evaluating the underlying values. Nonetheless, rural roots are often a source of memories and pride. They are among the best predictors of the environmental concerns and behavior of city adults. People's love of nature tends to persist despite large changes in other values and can be sustained by spending time with friends and kids in nature (Chapter 6).

Through surveys in many types of neighborhoods and communities, Joan Nassauer discovered that people who care for nature in their garden, farm, or community tend to inspire similar behavior among their neighbors, creating a halo of environmental stewardship.[18] Neighbors consider people with well-kept gardens, homes, farms, or businesses to be honorable people and responsible citizens whose behavior they want to emulate—even though they may not know these people personally. I knew that people often copy the bad consumption habits of neighbors (Chapter 6), but Joan's work was the first to convince me that good environmental behavior can inspire an environmental ethic among others and empower neighbors to be good stewards.

Education

The sharing of values among people is inseparable from education that occurs outside the classroom (**informal education**). People learn informally about the values and skills of people they admire. In this way, informal education empowers people to shape their own lives and therefore their capacity to meet their personal, spiritual, and societal goals (Figure 5.1).

Formal classroom education is also important to empowerment. La'ona DeWilde's **formal education** in Alaska grew from informal roots. She was home-schooled by her parents, who taught her not only traditional sewing, beadwork, hunting, and trapping but also how to read and learn from books. After completing high school, La'ona was unsure what to do next. A family friend persuaded her to continue her formal education at the University of Alaska. She was the top student in my ecosystem ecology class, both because she was smart and because the course built on the ecological intuition she developed during her childhood in the woods. Her summer job as a wildland firefighter built on and reinforced these same skills.

La'ona's master's thesis documented the effects of human ignitions and climate change on wildland fire in Alaska. This work integrated her informal education in the woods and as a firefighter with computer skills and knowledge about climate change that she learned at the university. As in La'ona's case, formal education is most useful when it complements informal education by providing skills that are less deeply embedded in culture. These often include computer skills, disease prevention, language skills, and familiarity with the many rules and opportunities of the modern world.

Despite the importance of informal education, formal education matters in today's world. Other than parental income, level of formal education is the strongest predictor of a person's potential upward economic mobility, especially for people with a low income. In the United States, a high-school degree, college degree, and post-college education contribute incrementally to upward economic mobility, whereas people with little formal education have fewer job opportunities, lower pay, and less job security.[19] **Median household income** is one way to assess this relationship: half of the households earn more than this amount, and half earn less. In 2017, the median household income of a high school graduate in the United States was only half that of a college graduate. Unfortunately, children in poor families often can't afford higher education and must spend more time working to support their families rather than studying. These constraints contribute to cycles of poverty. With adequate government support for education, this vicious cycle could be broken. Sadly, voters often elect politicians who seek to cut funding for education in the name of reducing government spending, thus reducing opportunities for youth in their communities to thrive.

Through both informal and formal education, people learn how to learn. People who have learned the value of education in their youth encourage a positive attitude about learning in their children and grandchildren, thus helping sustain over the long run the capacity of people to evaluate and meet their needs.

Cultural and Social Rules

People's values and knowledge play out in the social relationships they have with one another and the rules (institutions) that structure these interactions. These relationships are an important source of well-being. In 1992, Mimi traveled to Yakutsk, Russia, to teach adult classes in business English. With the help of a friend, Mimi found housing with a widow whose two daughters had recently married and moved to Moscow. After the economic collapse of the Soviet Union in 1990, both food and jobs that paid a living wage were hard to find, so Rosa was glad of the extra income.

Mimi and Rosa became good friends, overcoming differences of age, language, and culture. When Mimi returned to Yakutsk for a second teaching stint the next year, she stayed with Rosa again. By this time, Rosa's daughter Olya, son-in-law Aliosha, and new grandson Ilia had moved back to Yakutsk because they couldn't afford to live on their own in Moscow. Rosa's small apartment was the hub of family life. During the day Olya knitted things to sell on the street, and at night she worked as a professional dancer for the state dance company. Aliosha couldn't find a job in his engineering specialty. However, part-time odd jobs helped him deal with family issues of childcare, shopping, and various crises of repairs and shortages. For many families, rural relatives were a critical lifeline to potatoes, cabbage, fish, and other staples. Family ties and friendships spread risks and opportunities among people who cared about one another and helped people cope from one day, year, or generation to the next.

Empowerment and people's belief that they can change their own lives (**self-efficacy**) depend on a mix of values and of skills obtained through education. Self-efficacy has at least two components—knowledge of how to accomplish something, which comes from education, and a person's perseverance and passion to address long-term goals ("**grit**"). Grit contributes substantially to people's success, regardless of educational background.[20] Aliosha's perseverance helped keep his family afloat when he had no job. When his family moved into Rosa's apartment in Yakutsk, he sat in on Mimi's English classes in his spare time and worked part time as the local representative of an equipment importer. As his English skills improved and he gained experience in dealing with shippers, customs officials, and customers, his role in the company expanded, allowing him to make better use of his training in engineering. The self-confidence that emerged from these experiences eventually enabled him to start his own company. In general, empowerment depends on the capacity of people to do things—a consequence of available resources, education, perseverance, health, and culturally defined relationships with other people.

Like Rosa's family, almost everyone experiences hard times. More than half of adult Americans spend at least one year in poverty (Chapter 6). People experience poverty for many reasons, including divorce, unexpected pregnancy, death of a parent, injury, loss or inability to find a job, a natural disaster, economic disruption, bad decisions, or just bad luck. These challenges cause both economic and psychological hardship. At such times, social relationships with family, friends, and community are critical to people's well-being, as in Rosa's family.

Mimi's experience and that of many others have etched in my mind the importance of social networks that I seldom think about but that are critical to the continuity of community well-being: medical, police, security, and other emergency services; insurance of many types; church and counseling services; food banks; colleagues; community groups—in addition to family and friends. Some of these networks, like churches and community celebrations, are largely volunteer efforts embedded in culture; others are government-supported, and still others spring up spontaneously when the need arises. The social cohesion that develops from community networks can be fostered by almost any activity that brings people together, often as volunteers, to address common concerns: soup kitchens, community events, stream restoration projects, church groups, reading or music groups, garden clubs, and many other activities that foster trust and mutual under-standing (Chapter 8).

The bottom line is that good social relations with family, friends, and community help sustain the well-being of both individuals and communities. The stress that poverty places on social relations can be alleviated by both government and community programs. Volunteer efforts by individuals and community groups bring the added value of social relationships that are critical to building trust and well-being in any community.

Empowerment also depends on external barriers imposed by political and socioeconomic realities that people cannot easily overcome on their own. Fostering empowerment therefore requires a two-pronged strategy—first, to enhance people's skills, capacity to act, and belief that they can make a difference and, second, to understand, work effectively within, and sometimes challenge the power structures that might otherwise prevent people from using their capacities to accomplish personal goals (Chapter 9).

Environmental and Human Health

Modern medicine has made incredible strides in reducing the occurrence and impacts of disease. However, as my granddaughter Adele pointed out to me (Prologue), good health also requires a healthy environment. Surprisingly, Maslow didn't include a healthy environment among his list of fundamental human needs—perhaps because he focused on the immediate determinants (outcomes) of well-being rather than the factors that sustain these conditions over the long run. Humanity is embedded in Earth's biosphere.[21] When ecosystems are degraded by pollution or other human impacts, they are less able to meet long-term human needs for food, clean water, protection from disease, and other services that nature provides to society (Chapter 3). In this way, environmental degradation reduces, over the long run, the potential of people to lead healthy satisfying lives.

Many of today's most serious health threats emerge from changes in human interactions with the environment[22]:

- deforestation and water projects have created new habitat for malaria-carrying mosquitoes and other disease vectors in close proximity to people
- deforestation, bushmeat harvest, and intensive livestock operations have transferred animal-infecting pathogens such as COVID-19, HIV, and West Nile virus to people

- exploding deer and tick populations, resulting from predator removal in many parts of the United States, have increased the incidence of Lyme disease
- exposure to secondhand smoke and other environmental contaminants has increased the risks of asthma and cancer
- expanded use of antibiotics to control disease in stockyards and fish farms has increased the frequency with which antibiotic-resistant bacteria evolve

Improvement in an ecosystem service doesn't always benefit vulnerable segments of society. Expansion of intensive agriculture in developing nations is intended, in part, to meet people's food needs—to increase their **food security**. However, agricultural intensification provides the greatest benefits to well-to-do landowners who can afford to buy the necessary equipment, fertilizers, and pesticides. Poor people can be harmed by these programs if they lose access to their lands or can't afford to buy food produced by intensive agriculture.[23] Similarly, some forestry programs enhance the income of large landowners and illegal loggers but eliminate or degrade community forests on which poorer people depend. African conservation programs sometimes displace or exclude local people from game parks in order to develop tourism but may eliminate livelihoods of entire communities that traditionally depended on these lands. Communities and policymakers should consider who wins and who loses when there are trade-offs that enhance some ecosystem services but reduce services needed by others.[24]

Technology and Economy

Human ingenuity and technology have created a smorgasbord of products to satisfy human needs and desires. In Kerala, southwest India, cell phones have created important new opportunities for people.[25] As long as anyone could remember, sardine fishers never knew at the beginning of the day whether they would come home with a large or a small catch. In some ways fishing success didn't matter. If people caught few fish, the price was reasonable, but they had few fish to sell. If the catch was good, every boat brought home lots of sardines, prices were low, and unsold fish had to be thrown away. The introduction of cell phones in 1997 changed all that. Fishers could compare sardine prices among neighboring ports and arrange a sale where the prices were best. Cell phones raised fishers' profits by 8% and reduced fish prices to local residents by 4%. Fewer fish were thrown away. The cell phones paid for themselves within a year or two, and everyone was better off. Similarly, cell phone apps enable Maasai herders to find water during droughts.[26] Cell phones are now used just as widely in sub-Saharan Africa as in the United States. They improve access to information in a region where conditions are uncertain, landlines are rare, and transportation or other forms of communication are time-consuming and undependable.

Technology that directly improves long-term economic opportunities for poor people provides a direct link to their well-being. In contrast, economic indicators of national wealth, such as GDP, are more tightly tied to the economic success of people who are well off. Increasing GDP trickles up to the wealthy but may actually undermine the ecosystem services that provide the greatest benefit to poor people. Perhaps that is why Maslow didn't include economic prosperity in his list of fundamental human needs.

There are so many ways in which technology has changed people's lives—in many cases clearly enhancing their well-being. Plows and high-yield crop varieties enhance food production, water-purification systems provide safe water, clothes and shelter protect people from bad weather, fuels and electricity provide convenient energy, and mechanized transport moves people and goods to places where they are needed. The well-being of many people depends on products like these, so the economic and technological capacity of a society to produce or buy these products enhances the well-being of its citizens. The utility of these and other products depends strongly, of course, on local context. Fishermen don't use plows, but they benefit directly from sonar that detects fish or reefs.

The downside of effective technology is that it increases the ease with which people find and extract resources. Sonar has contributed to marine overfishing by increasing the efficiency with which boats find and catch fish. Massive earth-moving machines enable coal companies to remove entire mountaintops and push them into valleys where people used to live and get their water. Environmental and ecological safeguards must therefore evolve in concert with changing technology.

Technology also creates products that have only tenuous connections to Maslow's fundamental elements of well-being. Video games teach children manual dexterity but detract from their time and interest in doing things that connect them more directly with nature,[27] so access to more manufactured products doesn't necessarily translate into greater well-being and indeed can be a cause of environmental degradation (Chapter 3).

In summary, many factors interact to sustain human well-being over the long run. Some of them, like formal education and empowerment, require actions of government. Others, like technology and economy, are fostered by business. All of these factors depend on the health of the environment and on the actions of individual people, either working on their own behalf through self-empowerment or as collaborative efforts for the benefit of society. Individual actions make a difference.

What Can We Do?

People everywhere have similar fundamental needs, but their relative importance varies with age, cultures, and segments of society. A person can improve her or his own well-being and that of the community and society in many ways:

- **Know what different groups of people need for a more satisfying life.** These probably include aspects of environmental and material conditions, social relationships, and empowerment to meet their goals.
- **Explore strategies with others to meet these needs.** Discussions with family, friends, and social networks, such as church groups, can foster the empathy needed to raise concern about the well-being of others, including vulnerable people like children, the elderly, and the poor.
- **Join efforts to improve the well-being of less fortunate people in the community.** This may include advocating for government programs that assist vulnerable people or volunteering in community programs to help them directly.

The next question is: *If we know, in general, how to foster the health of ecosystems (Chapter 2) and society (Chapter 5), what specific actions would have the greatest positive impact on the health of both?*

SECTION 2
GRASSROOTS ACTIONS

6

Individual Actions

What Can I Do?

The essence of stewardship is caring for nature's house. This is a daunting prospect because each of us in 2020 is only one 7.7-billionth of Earth's population. How can one person possibly make a difference? The most pleasant starting point is to recognize, celebrate, and reinforce the inspiration and joy that come from being part of nature—to flourish as part of nature's household. A necessary complement is for each person to reduce his or her detrimental impacts on nature so that everyone can enjoy its exuberance (Chapter 3). In these two ways each person can make a personal commitment to shaping a positive future for humanity and our planet. This chapter describes this foundational element of my four-tiered stewardship strategy.

Connecting with Sense of Place

When I was about 10, my family spent the summer at my grandmother's house in western Washington. In the afternoons, my father and a local friend who knew the area like the back of his hand often took me on their trout-fishing expeditions. We hiked to streams through cool shade beneath massive Douglas fir trees or scrambled over slash where the forest had been logged. On special days, we worked our way down the White Salmon Canyon to places where fishermen seldom went. I remember the anticipation of discovering the perfect pool, wondering if it held a monster trout that I could tempt to bite—or perhaps even catch. The yellow-jacket nests, worry of falling in, and long walks back to the car that bothered me at the time have largely faded from memory; but I will never forget the fishing or the woods. Even though my time spent fishing, huckleberry picking, hiking, and camping with family probably accounted for less than 1% of my time growing up, those are the clearest of my childhood memories and undoubtedly nudged me toward my career in ecology. Teaching ecology gave me the chance to share nature with students and hopefully plant or nourish a seed for their future connections to the nature that surrounds them. These experiences also shaped my father's actions. After he no longer fished, he spent his retirement years supporting efforts to conserve the lands where we fished together.

My grandfather was a city boy who grew up in New York City, as had two generations before him. How did my father, with his deep urban heritage, ever become interested in spending time in nature? Shortly before his death, I rummaged through family photos, looking for hints of stories I might ask him about. Deep in boxes of family history, I discovered pictures of my grandfather camping and fishing with his father in Maine, while his mother painted landscapes. Although they probably didn't spend much time in non-urban nature, the photo albums showed that these were seminal experiences that they chose to record for posterity. This habit of spending family time in nature cascaded down through at least six generations to my grandchildren, infused by parents and passed on to kids (Figure 6.1).

I know more directly how my son Keith and his kids interact with urban nature in their daily life in Cardiff, Wales. Most mornings Keith jogs through

Figure 6.1 Six generations of time spent in nature, from my great-grandfather and his son, who grew up in New York City but loved fishing, to my grandchildren, Adele and Emile, who are also city people who love to play and hike in nature.

the park along the River Taff, accompanied by a chorus of birds and the occasional fox or hedgehog. Each season has its own hues and special places—the gray winter mist rising off the river, the subdued autumn reds and yellows, and the audacious yellow daffodils of spring. This urban nature insulates him from the cacophony of life's daily details and gives him the luxury to think about just one thing. At other times, he thinks of nothing in particular and simply absorbs whatever nature has to offer. This link with nature motivates family projects in Keith's postage-stamp backyard garden, where new plants squeeze into nooks left behind by last summer's annuals, and bees share Keith's appreciation for gaudy pink and purple fuchsias. The three birch trees that Keith planted subtly remind him of the Alaskan birch forests where he learned to ski. Nature spills indoors where plants settle onto window sills and peek over shoulders to calmly watch the morning breakfast, afternoon cello practice, or Keith's preparation for the next day's lecture. Keith's kids relate to urban nature differently than he does. They use trees as goal posts to score the winning goal, look down at ant-sized people from tall trees, or hide from grandparents in dense jungles of camellias. Sage, rosemary, and thyme link the backyard garden with Friday night pizza.

Fortunately, the process of connecting with nature is one of the easiest and most pleasant steps toward stewardship I can imagine. It's a starting point for sharing a stewardship ethic across generations.

Sense of place describes how people attach to, depend on, and shape places, as well as the meanings, values, and feelings that they associate with a place.[1] It develops by spending meaningful time in a place with family or friends; learning about the place through personal experiences, stories, and writings; or pursuing a livelihood rooted in that place. Time spent in a place, within and across generations, is the best predictor of people's attachment to places, even though some recent arrivals attach quickly to new places and others remain more rooted to their place of origin. Attachment to places reflects both *dependence* on places for livelihoods and the *identity* that people derive from places they care about.

Place dependence and identity build on one another. People who are deeply attached to a place frequently choose place-based livelihoods, such as farming, ranching, or local tourism, or move to valued places after they retire. These choices are often motivated as much by their identity that emerges from a valued place as by the profit they expect the place to provide. In addition, people with place-based livelihoods often come to care deeply about their place. Valued places are a source of personal and community identity and of the empathy that motivates caring for these places.[2] This caring is especially deep for people whose families have been rooted in a place for generations.

It isn't always easy to make ends meet when living off the land. Sometimes the economic difficulties of place-based livelihoods, such as farming or ranching, can be softened by government subsidies and incentives, such as conservation easements, that reduce taxes in return for a commitment to sustain traditional uses of the land. Some families supplement their place-based income with other jobs, like teaching, road maintenance, or a small business, that draw from a different segment of the local economy. Economic diversification provides ways that individuals can sustain their place-based livelihoods despite changing economic realities.

Even if people care about places, they may feel no responsibility to take actions to protect them.[3] This disengagement is particularly likely if they are satisfied with the current status of their place or feel powerless to make a difference. People are most likely to act on behalf of beloved places when conditions are getting worse—for example, through pollution or local development pressures. Sense of place therefore seems to be a *latent* potential of people to act on behalf of places they care about, should the need arise. The challenge is to activate that potential so that it becomes part of life's everyday thoughts and habits and a motivation for more concerted action when valued places are threatened.

Empathy can play a key role in motivating action in support of nature and the people who depend on it—both locally and in other places.[4] Empathy involves three interrelated steps. First, people must understand the impacts of environmental changes on nature. Second, they must feel an emotional connection to the place or people who have been affected. Third, this empathy must motivate them to take collective action on behalf of threatened people and places. Empathy that builds on this combination of knowledge, emotional connection, and action sometimes engages people who have only a loose connection to the places that have been affected. Floods in southern England in 2013–2014 engaged people from near and far to raise funds and volunteer to help. In this way, empathy expanded the geographic scale over which caring led to action.[5] Pathways that build on empathy can engage people in responding to climate change in ways that intellectual arguments would never touch. Empathy links the heart to the mind.

There are many ways to foster empathy.[6] An *understanding* of the changes that affect nature and people's lives comes from observing, reading, discussing with others, and pondering the impacts. The *emotional connections* of empathy can arise and grow from videos, art, or stories about people who are harmed through no fault of their own. Role-playing exercises are particularly effective because they put people in the virtual shoes of others whom they do not know to experience the emotional trauma of vulnerability. These role-playing exercises are similar to scenarios developed in business, science, and politics to explore plausible outcomes of changes that are occurring (Chapter 7). Social networks and community groups often trigger the move from empathy to action. All three components of empathy are essential if empathy is to be effective. Understanding without emotional connection is sterile. Passion without understanding bypasses critical thinking and can trigger knee-jerk reactions with unintended consequences. Empathy without action creates more frustration than solutions.

Rural kids sometimes get bored with local social life and wish they had greater access to city glitz. Lifestyle choices and economic opportunities are some of the factors that fuel rural migration to cities throughout the world. My mother-in-law was glad to escape the Massachusetts farm where she grew up, but she never lost her love of gardening. My wife Mimi spent childhood summers with grandparents and cousins on that same farm. This is still the place to which she traces her roots of loving nature. We raised our family in Alaskan nature, and our kids and grandkids still identify with those roots. In general, social scientists find that people who grow up in rural areas are more likely to spend time in nature as adults and to recognize the value of nature to their psychological and social health.

An increasing proportion of the world's population—more than half— now lives in cities. Rural roots therefore become increasingly distant from the place attachment of city residents. As a growing demographic majority, city residents will play an increasing role in the election of leaders and choices of local, national, and global policies that shape the future relationship between people and the rest of nature. Opportunities for city people to experience and enjoy nature and to celebrate its values are therefore particularly important. A connection to urban nature is not too difficult to establish and sustain.

Many cities recognize the value of urban nature for attracting visitors, as well as for the psychological and social health of their residents. Urban renewal of aging infrastructure enables cities to reinvent their landscapes so that residents can connect more easily with nature.

Reinvention of Providence and Other Cities

Providence, Rhode Island, was one of the first American cities to industrialize.[7] Its three rivers were initially crucial for transporting goods to and from downtown Providence. However, these rivers eventually became a nuisance and were bridged over and routed beneath highways that cut up the city center.

When textile and other industries went into decline in the mid-20th century, the city center was clogged by an underused railroad yard and the intersection of interstate highways. Through a series of renewal projects, the city moved its railways and highways and uncovered and restored its rivers. Parks were built along the riverbanks. An annual celebration of water and fire brought thousands of residents and visitors to the parks in the city center. This restoration of urban nature was critical to the revitalization and development of a new identity for downtown Providence.

Similarly, Baltimore, Maryland, rejuvenated its harbor; San Antonio, Texas, its riverfront; Portland, Oregon, its urban center; and New York City its abandoned elevated rail tracks (High Line). About half of a city's infrastructure is replaced each half-century. That's a lot of urban renewal every year and provides ongoing opportunities to integrate city life with nature. This makes the city more attractive to both residents and visitors.[8]

Giving up on Freeways

Americans' love affair with their cars has transformed many cities into spiderwebs of freeways. Los Angeles, California, is a classic case of the resulting imprisonment of people in cars and gridlock traffic. It is almost impossible to walk from place to place in downtown Los Angeles or even to navigate by public transportation. And yet, building more freeways just attracts more cars (**induced traffic demand**). The expansion of Interstate 405 through Los Angeles only made rush-hour congestion worse, so the city decided not to add lanes to another freeway.[9] London, Stockholm, and many Italian cities have reduced inner-city congestion by imposing fees or fines on people who drive in the city center. These fees support improved public transportation and bike paths in London and have sharply reduced pollution and children's asthma attacks in Stockholm. There is increasing pressure in many cities to eliminate rather than expand urban freeways as a path to making cities more livable.

Greening of Detroit

The Greening of Detroit is a grassroots effort to transform Detroit, Michigan, from a Rust Belt casualty to a vibrant green city.[10] Each year the program recruits about 4,000 volunteers to plant trees, pick up litter, work in urban gardens, and educate youth about the natural environment

(Figure 6.2). Most cities and towns have volunteer groups that provide opportunities for individuals of all ages to experience nature, while
improving it for others. In another grassroots effort, the Trust for Public
Land seeks to create a park within a 10-minute walk of every city resident

Figure 6.2 "Stop and smell the flowers" say volunteers at Greening of Detroit (top),
while Roosevelt University students remove invasive plant species at Eden Place Nature
Center in Chicago (bottom).

in the United States.[11] Initiatives like these enable volunteers to enhance human interaction with nature for both themselves and others.

Building Nature in Chicago

Fuller Park has been called one of Chicago's worst neighborhoods, based on almost every statistic. In 2013, this neighborhood led the city in poverty, distrust in neighbors, unemployment, **food insecurity** (too little affordable nutritious food), violent crime, and lead contamination. Sixty percent of the area was abandoned lots.

Community leaders Michael and Amelia Howard decided to change this. In 1997, they acquired the deed to a three-acre abandoned lot across the street from their house.[12] The lot had become an illegal dumpsite with mounds of waste up to two stories high. With the help of volunteers, they removed over 200 tons of waste from the site and replaced it with topsoil (Figure 6.2). They named this new park Eden Place. They built a reflecting pond as a place of respite from urban surroundings and restored a woodland around the "mighty oak" that had survived decades of urban blight. In 2004, only 7 years after the project began, the US Environmental Protection Agency awarded Eden Place the Conservation and Native Landscaping Award for its use of native plants and animals to contribute to the city's biodiversity. In 2012, 14,000 people (half of them schoolchildren) visited Eden Place. First Lady Michelle Obama and many other national leaders have called Eden Place an environmental success story. All this began with the efforts of two individuals—Michael and Amelia Howard.

I met Mila Kellen Marshall when she was a graduate student in the Ecological Society's SEEDS (Strategies for Ecology Education, Diversity and Sustainability) program to promote diversity in the field of ecology. Mila stood out as a leader among the many excellent SEEDS students. She was articulate and passionate about environmental justice and exuded optimism about what students in ecology can accomplish. She carried that energy and leadership into her work as research director at the Eden Place Nature Center.[13] Here, she mentored student volunteers and interns and organized the annual URBAANE (Urban Resolutions Bridging African Americans to Natural Environments) conference.[14] Her vision for connecting people with nature is amazing. "We don't want to be the Rust Belt. That's not sustainable.... We are a city of amazing nature and nature experts.... Reconnecting to the environment can position each of us to be a part of the solution and not the problem.

Acting local is easier than you think, and we can be a part of the legacy of ensuring a healthy Chicago for generations to come." Mila connects people with nature in ways that matter to them. Currently, as a graduate-student researcher at the University of Illinois Chicago, she studies the connections between race, space, and environmental quality for understanding sustainable food systems.

There are so many ways to immerse kids in nature—hiking or camping in mountains or forests, swimming or collecting seashells along the beach, spending time in summer camp, family gardens, parks, or zoos. Nonprofit groups are constantly inventing new ways to connect people with nature. Interfaith Power & Light offers seeds for a "pizza garden"—all the herbs and vegetables needed for pizza toppings.[15] Immersing kids in nature is perhaps the most enjoyable way to foster intergenerational respect and care for nature and to build a foundation for stewardship. And besides, it's fun and strengthens each person's relationships with friends and kids!

There are, of course, many people who have no explicit interest in nature or pay little attention to it. Some people prefer to spend an afternoon in a shopping mall or playing video games than walking in a park.[16] However, everyone depends on nature and experiences less psychological stress when nature is part of their environment, whether they are consciously aware of it or not (Chapter 3). There is a nearly universal preference for landscapes that contain nature.[17] Attachment to nature is part of our evolutionary heritage.[18] However, there is a shifting baseline—the less nature we experience, the less nature we seek to sustain in the future.[19] This is why it is so important to incorporate nature into the design and redevelopment of cities[20] and to expose kids to as many kinds of nature as possible—whether it is climbing city trees or walking in wilderness.

Impacts of Consumption on Nature

Society's impact on the planet is a direct result of our consumption of resources. Impacts depend on the food we eat, the infrastructure we build, the cars and electronic gadgets we buy, and the energy used to make and use these products and to move from place to place. Resource use is obviously essential in the modern world, so how do we make consumption choices that reduce our impact but still maintain a satisfying life?

Human impact on the environment wasn't always so heavy. Since World War II, society's consumption of resources and environmental impacts have accelerated tremendously (Figure 6.3).[21] Out-of-control consumerism is

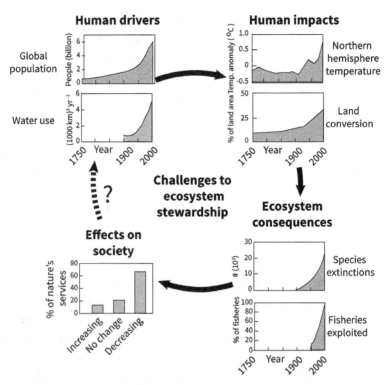

Figure 6.3 Challenges to stewardship result from human impacts on nature: increases in human population and consumption per person lead to changes in land use and climate, which affect the functioning of ecosystems. This, in turn, affects the benefits that society receives from ecosystems (Chapter 3). The greatest uncertainty is whether humanity can recognize these causal links and reduce the resource impacts of its consumption so as to reverse this vicious cycle.

not an American or developed-world tradition or economic necessity—it characterizes only 70 short years of inattention to society's growing impact on the environment. That's a fixable problem.

Although environmentalists often blame environmental damage on companies that extract fossil fuels, mine minerals, and clear land for agriculture, these activities are profitable largely because of public demand for products that we consume (Figure 6.4).[22] So society bears substantial responsibility for resource-extraction activities and associated pollution. Some of this consumption increase reflects the 4-fold increase in human population during the 20th century, but most of it results from increases

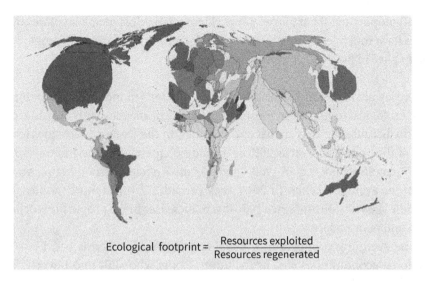

Figure 6.4 Ecological footprint of the consumption of renewable resources by each nation. Resource consumption by a country is calculated as the land area whose production is appropriated for human use plus imports from other countries minus exports, expressed in units of global hectares per person. Each country is stretched or shrunk to show the proportion of the global ecological footprint that it contributes.

in consumption per person, both by individuals and by government and business on our behalf. An individual in a developed nation consumes 32 times more resources on average than a person in the developing world.[23] Citizens of developed nations can therefore be 32 times more effective at reducing human impact on the global environment through well-considered consumption choices. In other words, they are poised to have the largest *positive* effect on the global environment by consuming more wisely.

Household income is a strong predictor of household consumption within and among countries.[24] China, India, and other densely populated nations with a rapidly growing middle class now also contribute substantially to global declines in ecosystem services as people seek to copy the consumption patterns of developed nations.[25] Now is therefore a crucial time for people in developed nations to set a better example of ways that less resource-intensive consumption can still provide a satisfying life.

In contrast, developing nations with large proportions of their populations in poverty, such as most of Africa and parts of South America, consume only a small proportion of Earth's resources (Figure 6.4). Most people in these countries have much less scope to reduce their consumption than do people who do not live in poverty.

There are also substantial variations within nations in income, consumption, and ecological footprint. In most nations, these disparities are increasing, primarily because the rich are becoming richer. In the United States, the richest 1% of the population earns 20% of the nation's pretax income and owns 35% of the country's assets.[26] In 1965, chief executive officers earned 20 times more than the average worker. In 2016, they earned 271 times more money! The middle class has shrunk since 1970 due to increases in the proportions of both rich and poor people.

The declining proportion of people in the middle class is bad news for national economies as well as for those people who sink into poverty. The middle class is an economic engine whose purchases have historically been a major driver of the US economy, so the greater disparity between rich and poor—and the shrinkage of the middle class—hurts most people. The United States has less income equality than 84% of the world's developed nations (according to the Organisation for Economic Co-operation and Development) and 83% of all nations.[27] In other words, there is greater income disparity between rich and poor people in the United States than in most other countries.

Most of us know people who have fallen on hard times. In fact, 60% of Americans between the ages of 25 and 60 have spent at least 1 year living in poverty.[28] My brother, for example, lost his health and job and moved into his car because he didn't have enough money to rent a room. Safety nets that support people during hard times usually come from friends and family who step up to provide support or from government, religious, or other charitable programs. I admire my friends, Lynn and Stan, for volunteering at the local soup kitchen. I also respect the local churches, charities, and government programs that finance these safety nets.

Safety nets work only because of support from individual people in the private and political spheres. Sadly, many people view government-supported safety nets as a tax drain on their finances rather than as ways to contribute to the lives of their less fortunate neighbors—and potentially themselves. Government safety nets are likely to be supported only if citizens demand this from their elected representatives.

These patterns suggest two complementary approaches to achieving a better match between resources used by society and societal well-being: (1) assure that the basic material needs of all people are met and (2) reduce the upward spiral of consumption by people who do not live in poverty by fostering components of happiness and life satisfaction that depend less on consuming natural resources.

Honorable Consumption

Consumption is not inherently bad. Within reason, consumption is necessary for individual survival and for vibrant local and national economies. Changing what, how, and where goods are produced can greatly reduce environmental impacts while maintaining material living standards. Producing food with less fertilizer and pesticides or producing energy from renewable sources rather than fossil fuels would greatly reduce human impacts on the planet, regardless of levels of consumption (Chapter 2).

In addition, people often make consumption choices that match their personal values. Some people grow vegetables and flowers rather than buying them at a store. After a long winter of eating rubbery tomatoes and tired salad greens from the store, Fairbanksans begin the gardening season with a vengeance, long before the snow melts. Carefully tended seedlings of tomatoes, basil, and squash displace the winter clutter on kitchen windowsills and living-room floors. Throughout the summer, people crowd the farmers' market to buy vegetables, homemade blueberry jam, fresh bread, or local pottery.

Many Fairbanksans volunteer at local community-supported agriculture (CSA) farms as a way to enjoy healthy food, support CSA farm projects, and socialize in the warm summer sun. People make these extra efforts because doing so matches their values of health, food quality, friendship, and community support. In autumn, the seasonal rhythm shifts to picking berries or mushrooms, canning salmon or garden vegetables, or hunting for moose and caribou.

As in most of the United States, Alaska's food stores and warehouses have space for only enough food to last a few days,[29] so goods at grocery stores thin out quickly when one of Alaska's three food-transporting ships breaks down. Supporting local agriculture, food storage, and foraging from nature therefore makes strategic sense as contributions to food security.

Some purchases and activities that are motivated by stewardship values benefit the community and planet:

- healthy local foods
- local mom-and-pop stores and restaurants that support the local economy

- products or investments that are certified for their favorable impact on the environment and workers
- fuel-efficient or battery-operated cars
- donations to causes we believe in—and the list goes on

Sometimes, however, people make consumption choices that do not align with their own professed personal values. I may purchase a more energy-consuming car, house, or vacation despite my alleged concern about the effects of energy consumption on climate. To the extent that people make these choices, we are **free riders**, who rely on others to bear the burden of reducing energy consumption. At the international level, countries like the United States that choose not to reduce their combustion of fossil fuels are also free riders that hope to benefit from the actions taken by other countries without contributing our share of the effort to solve climate problems.

Disconnects Between Consumption and Happiness

Among poor countries (those with an average per capita income less than $12,000—that is, the US-defined poverty threshold for a single individual), there is a strong correlation between a nation's wealth and the average self-assessed happiness of its citizens.[30] Similar correlations between income and happiness of poor people occur within nations.

I was surprised to learn, however, that happiness is unrelated to income or consumption for people who live above the poverty line (85%–90% of Americans).[31] As people become richer, they often aspire to even greater wealth, which, on average, reduces their happiness and overall satisfaction with their lives. Income has increased in many countries since World War II, but life satisfaction has not changed. As wealth increases, other factors, such as time spent with family and friends or in preparing one's children for their future, would contribute to happiness more strongly than does greater income and consumption.[32] If people chose to seek greater happiness and self-satisfaction rather than greater wealth, human pressure on the environment would decline, and most people would have more time and money to pursue their personal goals.

Why do most people above the poverty line seek to increase their income—and therefore consumption—when this does not increase their happiness or life satisfaction? Even billionaires generally seek to increase their wealth, even though they know their happiness is not limited by money. Some of us may never have thought about the disconnect between consumption and happiness.

For others, social status and power, which are often linked to wealth and consumption, may be stronger motivators for consumption choices than are happiness and well-being. People want to be like others whom they perceive to be more socially successful. Pressures to conform to the habits of others are especially strong for people who have recently moved to a new place, such as a city, where they seek to build new social ties. However, this is a slippery slope. If I buy a fancier car, clothes, or house in a more upscale neighborhood, I then see others who have still fancier cars, clothes, and houses, so my expectations escalate. It's our comparison with others around us rather than any specific level of material possessions that tends to define what people want. Children face the same escalation of social pressures because there are always other children who have things they don't have. At the other extreme, for people living in poverty, an increase in income may be essential to meet basic needs—including time spent with family and friends—and serve as a security buffer, in case their situation gets worse.

Celebrations provide a window into people's values. On the first American Thanksgiving Day, Euro-American colonists celebrated a bountiful harvest, thanks in part to the ecological approaches to farming they learned from local Native Americans. Many cultures celebrate in autumn the bounty that they harvest from nature. The day after Thanksgiving is now celebrated in the United States as Black Friday, which explicitly encourages unmitigated spending to maximize business profits—to keep business accounts in the black. This incongruous juxtaposition of celebrations—first of people's connection to nature, then of unmitigated spending—illustrates society's blindness to the impact of a consumptive lifestyle on nature. Although people are often motivated to shop at this time of year to bring happiness to others at Christmas, this motivation has been co-opted by advertisers as an opportunity to boost business profits through escalated consumption.

Advertising seeks to motivate people to buy products by informing potential buyers of the product's desirable features—a potentially useful role—and by tinkering with psychological dimensions of people's incentives to consume. Advertising often implicitly promises greater health, wealth, beauty, or power through television images of beautiful people and trucks in enviable social settings. Other ads promise health and security to the elderly.

Advertisers fuel the desire to consume. They may offer sales on specific items that may not be what a person really wants. The goal is to tempt people to buy something they hadn't thought of buying, perhaps something they didn't really need or a newer version of something they already have or to shift a person's allegiance to a new brand—perhaps a different, slightly more expensive car than their current one—even though the current car may still be in good condition. Once people have been enticed into a store, they often buy additional things that they never considered purchasing.

Sometimes businesses worry that people aren't replacing items often enough. One solution is **planned obsolescence**—building products that break after only a few years, perhaps because the new computer-controlled features are difficult to repair or parts are more fragile or no longer available from the dealer after a few years. Consumers are complicit in this racket if they make choices based on bells and whistles or appearance rather than service records.[33] Consumer Reports and other nonprofit groups provide performance information on alternative products for consumers who want to know about the durability of potential purchases. Buying based on performance can make a difference. When longer-lived Japanese cars were introduced into American markets in the 1960s and 1970s, American car manufacturers were forced to make cars that lasted longer.

Financial practices such as credit cards, student and auto loans, and mortgages encourage purchases without full consideration of the financial consequences. Increases in household debt now fuel US consumption more strongly than does employment.[34] Household debt increased in the United States 3.3-fold between 1947 and 2015. In 2011, average household debt in the United States was 112% of disposable income. In other words, the average household owes more money than it earns in a year! Households that cannot afford to pay the interest on their debt may be forced to sell their house or car, greatly reduce their spending, or—more likely—continue to accumulate debt and spend more of their income as interest on that debt. Financial insecurity may partially explain why happiness in the United States has not kept pace with rising income and consumption.

Less Consumptive Pathways to Well-Being

I've never met a perfect environmental citizen and certainly don't claim to be one. I consume more than necessary, and some of my choices don't favor society or the environment. But, like many people, I try to think about which choices matter and, within reason, I do the best I can. I admire people like Gandhi who lived a life of self-sacrifice, but that's not the life I've chosen. How can I reduce consumption and still lead a satisfying life? It's actually pretty easy. Here are a few examples.

Like everyone, I buy some unnecessary things, sometimes because I think I might want or need them in the future, sometimes as gifts for others, and sometimes for no good reason at all. Most of the stuff that fills my closets and shelves has not been used for at least 2 years: a closet full of seldom-used shirts, a dehumidifier that I used once 5 years ago, a lamp that I never got around to assembling, books that I intended to read and then forgot. How much of this stuff did I really need to buy? How many times have I bought several specialty items when a single general-purpose tool or pair of shoes might have done the job? How many times have I bought something that is bigger or fancier than I really needed?

Even for necessities like food, people often buy more than they need. People throw away some purchased food because they forget to use it while it is fresh. They could have bought less or waited until the following week to buy what they didn't immediately need. I was surprised to learn that about a third of US food production is thrown away, about 40% of this by individual consumers in their homes (Chapter 3).[35] Agricultural production that is thrown away benefits no one. But it has a massive environmental footprint when aggregated at the national or global scale and reduces the effectiveness of agriculture at meeting global food needs.

Most of us have some old things given to us by friends or relatives that we cherish because of their personal associations—the moth-eaten blanket that my wife's grandmother gave us after our wedding, which is still warmer than potential replacements from the store. I buy other things from friends or organizations I admire or because I know they are well made and will last a long time. Because I care about these special items, I usually keep them longer. The same thought process applies to gifts for others.

Other commercial purchases, like cell phones, water bottles, and T-shirts, mean less to me. I'm more likely to tire of them and replace them with something new, even though they still serve their purpose. But it doesn't need to be this way. I have a personal pride in maintaining and minimizing abuse to the used car I bought 18 years ago. Besides, I'm curious to know how long it will last before it needs major repairs. Perhaps by that time electric cars or driverless ride-share cars will be the norm, and I won't need to replace it.

Sharing and borrowing are natural activities in families and in close social networks where people trust one another. It's a logical way to avoid unnecessary purchases and clutter, while still accessing infrequently used items. Libraries are a public model for sharing and borrowing, where people act on an honor code to return books that they borrow—or risk losing their library privileges. Imagine the clutter if we bought and kept every book we ever intended to read! There are also commercial models for renting expensive items, such as cars, apartments, and equipment that are used only temporarily. These and other models of sharing, such as toy libraries, tool libraries, and time banking to share skills, are ways to access a variety of goods and services for short periods of time.

Most people with kids receive and give their friends hand-me-down outgrown clothes and toys or take them to charity shops. Cultures differ in the social acceptability of giving and receiving gifts from those who are not close friends. In many Native American communities, potlatches and other celebrations are formal times when people expect to give things to others— much like bringing food to share at a potluck meal. In other cultural settings, receiving gifts from those whom we do not know well bears a social stigma. Community and faith groups are often great starting points for innovative arrangements for sharing, borrowing, and giving because they build on a foundation of trust and caring among people with common interests.

Fairbanks, where I live, has no garbage or recycle collection in outlying areas. Instead, local government maintains "transfer stations," where people bring their garbage for centralized collection. One corner of this facility is reserved for useful, unwanted items, such as furniture, books, knickknacks, and appliances. Most of these are picked up by others who use or sell them. When I was a graduate student, places like this were the source of most of my furniture. I was shocked when I went to a community waste facility in California and was told I could not take away items that were still perfectly useable.

Most communities in the United States have facilities that recycle many of the materials that we throw away. My friend in Switzerland pays a surcharge for anything that doesn't fit into a small weekly bundle of non-recyclable garbage. This surcharge creates an economic incentive to avoid purchasing products with excessive packaging, which, in turn, puts pressure on companies to sell us only what we want. Taking reusable bags to the grocery store avoids the need for plastic bags that are thrown away as soon as a person reaches home.

My friend and research collaborator Sergey Zimov, who directs a research station in northeast Siberia, saves bottles, cans, broken-down vehicles— anything that is potentially useful. This "resource pile" substitutes for the local hardware store when something needs fixing or he begins a new experiment. Sergey and most people who live in remote Siberian and Alaskan villages know how to improvise with what they have. Recycling is more than a yuppie expectation. It's a necessity for getting things done with what is available.

The personal choices that most strongly influence household energy use are (from largest to smallest impact) family size, owning a car, flying on airplanes, using non-renewable energy, and choosing a meat-based instead of a plant-based diet (Figure 6.5).[36] Each person evaluates these deeply personal lifestyle choices differently and may or may not have options for alternative choices.

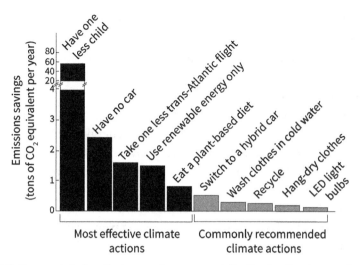

Figure 6.5 Personal choices that can reduce our contribution to climate warming in units of tons per year of CO_2-equivalent emissions—that is, emissions of all greenhouse gases converted to a common (CO_2) equivalent impact on climate warming.

For each of these choices there are intermediate options. Young millennial adults often marry at an older age than their parents (delaying the start of their family), which effectively reduces the rate at which family consumption and energy use increase. They may also eat less meat than their parents and use public transportation rather than owning a car. These lifestyle choices are made for many reasons, often unrelated to energy conservation.

Personal choices with more modest impact on energy use are those that most of us associate with energy conservation. These include replacing a typical car with a hybrid, washing clothes in cold water, recycling, hanging clothes to dry, and switching from incandescent to LED lightbulbs (Figure 6.5).[37]

Additional factors that strongly influence household energy use include the number and types of family cars, home size and energy efficiency, distance between home and work, and so forth (Appendix). Each of these factors offers a spectrum of options for reducing energy use.

- *Home heating and cooling.* Some houses require much more energy than others to heat or cool, depending on size, insulation, temperature settings, and climate. Energy audits pinpoint reasons for excessive heat exchange (such as poorly insulated windows or gaps in insulation), ways to regulate temperature more effectively (smart thermostats that account for daily and seasonal variation in temperature preference and electricity rates), and personal habits that use natural ventilation and passive solar heating (raising and lowering window shades on south-facing windows) to regulate temperature. In these ways, nature regulates the temperature of my home for about 7 months a year. This would be even easier in a climate that is warmer than Alaska's.

- *Electricity use.* Many books and websites list habits that reduce electricity use (Appendix). These include choosing energy-efficient appliances, unplugging freezers and turning off lights when not in use, and using less hot water. On the supply side, solar panels reduce energy needed from local utilities.
- *Fuel use.* Fuel efficiency can be maximized through choice of vehicles and a spectrum of driving habits, such as sharing trips, driving speed, regular vehicle maintenance, and maintaining air pressure of tires.

Are these actions enough? Definitely not, but they're a start; and they give me greater life satisfaction than if I choose not to do them. Many energy-saving habits require virtually no time, effort, or inconvenience. They are a bit like closing the door when leaving home—just part of daily routine. I often notice other people's energy-saving habits and consider adding them to my lifestyle. As a rough check on whether little habits matter, I looked back at records of my energy use a few years ago. My use of electricity, miles driven, and miles flown have declined about 50% without my noticing any inconvenience. Some of this reflects greater flexibility after retiring from my job, but much of it also reflects deliberate choices. On the other hand, I have made less progress in reducing fuel used to heat my home. This year I used savings from flying fewer air miles to buy a more energy-efficient furnace and am looking for other ways to improve the energy efficiency of my home. Similarly, climate scientist Peter Kalmas describes easy ways in which he and his family of four can live a pleasant life with only a tenth of the fossil fuels used by the average American household.[38] Reducing personal consumption is not that hard or unpleasant.

Trade-offs

Other consumption choices involve choices that have both advantages and disadvantages (**trade-offs**) that each person will evaluate differently. Most trade-off decisions provide a spectrum of choices, allowing people to adjust their choices to fit personal values and circumstances. This range of choices lets me try less consumptive options and personally assess their advantages and disadvantages. Here are a few examples.

Most reductions in unnecessary consumption and energy use are more of a mutual benefit (**synergy**) than a trade-off. The monetary savings reduce financial risks and debt and open opportunities for alternative pathways to happiness and well-being, such as less need to work long hours, more hours spent with family and friends, or greater investment in educational opportunities for my grandchildren. However, if the cost savings are spent on an additional airplane flight to a distant vacation spot, the environmental benefit of reducing consumption may be canceled out.

Commuting to work by bicycle has health and environmental benefits and time disadvantages compared to driving by car. Public infrastructure that is bicycle- and pedestrian-friendly or greater use of public transit also reduces people's energy use. Besides, riding a bus gives a person time to read a book. As with many trade-offs, intermediate choices, such as eating less meat or commuting by bicycle or walking on some days, provide an opportunity to test the fit of different trade-off options to personal lifestyles.

After graduate school, Mimi and I and our two toddlers moved to Alaska, where I began a job at the University of Alaska. Mimi also had good prospects for a teaching job based on her master's in teaching degree and the urgent need for more teachers in Alaska at the time. But we were reluctant to put our kids in daycare for their immediate future. Mimi chose to work at part-time jobs, some of which could be done at home, rather than becoming a full-time teacher. Since both of our jobs had flexible hours, we spent more time in family activities. Since we had less money than if we had two full-time jobs, we spent and consumed less.

Other people negotiate work arrangements that give them more time with family. Some businesses allow their employees to work 4-day rather than 5-day weeks and often find that 4-day workers work more efficiently and accomplish just as much because of greater satisfaction with their personal lives. Alternatively, other people choose to work overtime, which cuts more deeply into time with their family and friends. Some people telecommute, giving up the joys of traffic jams for time with family and friends. Options to work fewer hours are not available to everyone, and they usually involve a trade-off between a smaller income and greater opportunities to spend time pursuing other goals.

After living 16 years in Alaska, Mimi and I wanted to see what life was like in other places. We decided to move to the San Francisco Bay area and began looking for a place to live. Our love of Alaska's open spaces led us to look first at distant bedroom communities. However, the traffic we encountered as we looked for distant housing was horrendous. In the end, we purchased a small 1930s tract house within an easy bicycle ride from the campus where I worked.

Although we didn't like living cheek by jowl with the rest of humanity, our opportunity to spend free time hiking in nearby parks, playing Irish music in the local pub, and going to concerts was more satisfying than time spent commuting through gridlock traffic. Eventually urban life got the better of us, so we began looking for jobs elsewhere and eventually moved back to Alaska.

Government and the private sector also make massive consumption decisions such as the construction and maintenance of highways, factories, and power plants. Individual citizens influence these consumption decisions only indirectly in the messy world of politics and power (Chapter 9).

Unconscious Choices to Consume

Many consumption choices are unconscious. One of my favorite examples of this was a study conducted by a team of social psychologists.[39] They were asked by the US hotel industry to evaluate which messages would most effectively motivate guests to reuse their towels. The messages tested were as follows:

- "Please reuse your towels."
- "We will reduce your hotel bill by $5 for every day that you reuse your towels."
- "Please reuse your towels to help reduce impact on the environment."
- "75% of guests in this hotel reuse their towels. Please do the same."
- "Three of the last four guests in this room reused their towels. Please do the same."

The research showed no difference in reuse of towels among the first three groups of guests but greater reuse of towels by people who were informed that other hotel guests did so, especially guests that had previously used the same room. Clearly, conformity was more important than monetary or conservation incentives, and the more similar the conforming group, the more effective was the message in eliciting the desired behavior.[40] When this experiment was repeated in German hotels, both environmental and conformity messages increased towel reuse,[41] suggesting that conformity was effective in both situations but that Americans were less receptive to the environmental message than were Germans. Conformity thus sometimes shapes consumption choices more powerfully than do conscious rational choices about what we want or need. The German towel experiment shows that conformity depends on context. It's more important in some situations than others.

Groups shape who we think we are or want to be (our sense of **identity**). Most of us identify with several groups of people, including our nation, gender, race, profession, political party, neighbors, and friends. In the process, we often conform to the attitudes, rules of behavior, and choices that we think characterize "our group"—its **social norms**.[42] Once we identify with a particular group, our identity and self-esteem tend to become associated with membership in that group. Students identify with other students, parents with other parents, Catholics with other Catholics, environmentalists with other environmentalists, fans of a particular football team with other fans of the same team, and so forth. Members of the same group often have common interests and concerns, as well as expectations of similar attitudes and behavior about issues that are important to the group. On the other hand, distinct groups differ in some of their norms, choices, and behavior, for example, between career-focused people and those whose lives are more centered on their leisure time or on raising children, between miners and conservationists, between business owners and blue-collar workers, between large- and small-scale farmers.

Identification with social groups is a normal process that helps each of us decide how to behave in a complex and confusing world. Social groups are valuable because they provide friendship, respect, and trusted sources of information about issues that we don't fully understand or want to make the effort to learn about. Therefore, our choice of social groups and our interactions with people in these groups tend to lead to common perspectives on issues and conformity to the predominant values, attitudes, choices, and behaviors of people within the group. We may be less likely to show opinions, choices, or behavior that we think will cause disapproval by our groups. Social groups lead us to make some choices without consciously thinking about them.

Conformity within social groups provides both opportunities and barriers to meeting society's needs. It is good that most people conform to general traffic rules—driving on the right-hand side of the road in the United States but the left-hand side of the road in the United Kingdom. These habits that we don't think about allow us to focus attention on less predictable issues, such as the behavior of pedestrians or the color of traffic signals. Similarly, when Mimi and I moved to Alaska and couldn't afford to buy a house, we built one ourselves, in part because several friends and people we respected had done this. If we had moved elsewhere and formed friendships with people who had never built a house, I doubt we would have considered this option. Other decisions about family size and timing are also influenced by social groups. Thus, both big and small decisions can be strongly influenced by social conformity.

Just as people may tend to unconsciously conform to the choices and actions of others in their social group, their personal choices and actions influence their friends and neighbors, creating an opportunity for wise consumption choices to spread beyond the individuals who make them.[43] As shown by the use of towels in hotels or social disapproval of smoking, conformity has its greatest impact on people whom we view as members of our own group—our family, neighbors, and other groups we belong or aspire to. The larger the proportion of people who make wise consumption choices, the more likely that these choices will spread. Behavior that is highly visible, such as recycling of wastes, traveling to work by bicycle, or refraining from smoking, is especially likely to be adopted by our neighbors or social group.[44] Similarly, those who care for nature in their yard or farm are respected as good stewards and responsible citizens by their neighbors.[45] In these ways, groups of people with wise consumption patterns and care for nature become role models for others, and the pattern has the potential to spread.

In summary, choices by individuals to spend time in nature and to seek less consumptive pathways to life satisfaction are critical starting points for stewardship. Many guidebooks and websites supplement the strategies outlined in this chapter about specific choices and actions that are good for society and the environment (Appendix). Initially, people must pay attention to their consumption choices—assessing how well these choices align with their personal values. Once these behavior patterns become everyday habits, little effort is required to continue acting responsibly for the future of our planet. Behaviors that are good for the planet and contribute to our own life satisfaction are stewardship norms that have the potential to spread among friends and neighbors. Some people will, of course, make choices that do not reflect stewardship. The next task is to explore steps that communicate and foster the expansion of stewardship values and behavior beyond our group to the rest of society.

What Can We Do?

The actions of each individual are a fundamental starting point for stewardship. Constructive actions people can choose include the following:

- **Experience and celebrate the joys of nature.** This can be done in many ways and places—from cities to wilderness. Sharing the joy of nature with others—especially children—teaches them to appreciate and care for nature.

- **Understand how our consumption choices influence nature.** This understanding helps us prioritize decisions and actions that have the greatest positive influence on the environment and future generations.
- **Be proud of actions that minimize environmental impacts.** With time, these **honorable consumption** decisions become habitual and require little conscious thought. They also influence the behavior of others through spreading of social norms.
- **Reduce unnecessary purchases.** Pay particular attention to products and actions with the most negative environmental impacts.

The next question is: *How can effective communication foster the spread of sustainable choices?*

7

Dialogues for Solutions

Dialogues that respectfully communicate stewardship goals and behavior can shape the social norms and behavior of others. Communicating effectively, as described in this chapter, is the second element in my four-tiered stewardship strategy.

Developing Trust to Solve Problems

People are often skeptical of the motives of people they don't know. Over the years, Alaska Natives came to distrust many of the researchers who came to their communities.[1] In April 2009, Larry Merculieff, an Alaska Native leader, put this issue squarely in my face. I had invited him to give a talk about indigenous perspectives on sustainability at the University of Alaska Fairbanks.

"It's time you researchers stop exploiting native communities and do something useful for a change," Larry said, as he began his talk. "You typically decide what you want to study, parachute into a community without asking permission, do your study, leave, publish the results, and the community never even learns your findings. You need to turn that relationship around. Communities should decide what needs to be researched and who should be invited to do this work. The exploitative relationship between researchers and communities has got to stop."

Larry was right. His talk triggered a 2-year conversation among a dozen native leaders and university researchers. Together, we designed an *in-reach* program from native communities to people with technical expertise at the university and government agencies to address *native people's* visions for the future.[2] We knew this partnership would work only if it was grounded in mutual respect and a willingness to share responsibility and power. Native communities therefore decided the issues to be addressed, the research process, and the results to be shared. Our role at the university was to help find expertise that might be useful.

At Larry's suggestion, we called this the Community Partnership for Self-Reliance (CPS). Its goal was to empower communities to solve their own problems rather than increasing their dependence on government programs. I was dubious that this framework would resonate with communities, given their scarcity of jobs and high poverty level; but I trusted Larry's judgment. Larry and Patricia Cochran, executive director of the Alaska Native Science Commission, agreed to help us engage the first four communities that applied to join this partnership.

Igiugig, a Yupik village of 70 people in southwest Alaska, was the first community that CPS visited. As our plane approached the village, all I could see was snow. Eventually, I picked out the snow-covered landing strip, then the school, and 20–30 houses along the river. Like most communities in rural Alaska, Igiugig has no connection to the road system or electricity grid that serve larger cities, and most transport is by air. Christina, the environmental coordinator, met the plane. She knew Patricia from a stint in Anchorage as an intern with the Alaska Native Science Commission. They chatted and caught up on news about mutual friends.

The CPS visitors—Larry, Patricia, graduate student Becky Warren, and I—and tribal council members gradually assembled in the tribal office and introduced one another. AlexAnna, the tribal administrator, asked why we had come. It made no sense to her that the CPS visitors would fly all the way from Anchorage to explore a collaborative partnership without having an agenda of projects we wanted to push. Larry and Patricia explained that we wanted to learn what were the concerns of the community so that we could explore projects that *they* thought were important. AlexAnna and the other members of the tribal council trusted Patricia as a friend and Larry as a well-known advocate of native rights, so they rattled off a list of concerns—and Patricia, Larry, Becky, and I listened.

Later, at the community-wide meeting, residents wandered in and helped themselves to the fruit and cookies that Patricia had brought from Anchorage. After a prayer by a village elder, Larry launched into a story about how his elders had told him that things would get much worse before they got better. Communities would need to rely on traditions of sharing and working together to make it through the hard times. Prices in rural villages would continue to rise, and government agencies would cut back on programs that currently support villages. His message was a brutally blunt explanation of the CPS concept of self-reliance. I was glad it wasn't me trying to push that message, and I wondered how the community would respond.

One after another, people from the community stood up and described similar stories from their elders. They gave examples of problems they had solved on their own and described declines in government support services. Although Larry's story portrayed a tough future, it put the focus squarely on community empowerment to decide what was important and what to do about it.

The next morning Christina and AlexAnna showed our CPS group around the village so that we could see how the community was addressing problems they faced. Larry, Patricia, Becky, and I joined in the conversations, mainly asking questions about the projects they had done and their strategies for addressing remaining challenges. In most cases, they had a plan for moving forward, and we listened.

However, two unsolved issues came up repeatedly: the high cost of transporting diesel fuel (their main energy source for heating and electricity) by plane and threats to their legal rights to clean water. Fuel in Igiugig was more than twice as expensive as in larger communities along the road system. The river that provided their drinking water and salmon was being polluted by commercial hunting lodges. The lodge closest to the village threatened to sue the village if it was reported for dumping raw sewage into the river. These issues seemed poised for collaboration. I knew colleagues who were studying the integration of renewable energy into diesel power systems and others who worked with the state of Alaska on monitoring, protecting, and enforcing water quality. CPS became a matchmaker between the community, which had defined its two top concerns, and researchers and agencies with the expertise to address these concerns.

I'm now a firm believer in starting conversations about problem-solving with the people who are most strongly affected. Studies that take the time to involve their target audience produce results almost 10 times faster than studies that don't engage local residents.[3] After a couple of trips to Igiugig, I became good friends with AlexAnna, Christina, and their neighbors. When I first went there, local residents had never heard of me, so they had no reason to trust me as a university researcher. Patricia's friendship with people in both the village and the university established an initial bridge for trust-building. She was a trusted messenger.[4] Larry's message about self-reliance put the focus squarely on community-based concerns and solutions. Trust solidified as we engaged in conversations in which we each respected and learned from the expertise and experience of others. AlexAnna, Christina, Patricia, and Larry have become my go-to people whenever I want to understand Alaskan village perspectives on important issues. Besides, we have fun talking about stuff that has nothing to do with science or politics.

When I visit a village with someone who knows it well, our first day is always spent visiting friends, drinking tea, and catching up on local gossip—or perhaps helping out on a project that needs an extra pair of hands. Only after renewing and making new friendships do we begin talking about projects. Building trust in a village happens most easily by tagging along with friends who are already trusted by the community. This approach is similar to ways I've become involved in the stewardship efforts of other groups: other disciplines at my university, scientific societies, community action groups, faith groups. I never feel effective in contributing to efforts I care about unless I take time to listen and make friends, develop trust with members of the group, and learn their perspectives on how we might work together. Trust-building takes time and a willingness to listen, but it is time well spent. It saves time in the long run.[5]

As in Igiugig, trust-building is most effective when done jointly with people the audience already trusts—a trusted messenger. For example, Students for Carbon Dividends is a coalition of 100 college-student groups—primarily Republican groups—that are concerned about climate change.[6] They advocate for renewable energy and a carbon tax linked to a carbon dividend that would incentivize a transition away from use of fossil fuels (Chapter 8).[7] Young people are more likely than their parents to have taken a climate-science course, and young people have more to gain or lose from society's willingness to tackle climate change. Students for Carbon Dividends and other groups of young people, such as Young Conservatives for Energy Reform,[8] seek to take leadership in bridging the partisan political divide and fostering true dialogue about climate issues. Millennial Republicans born in 1981–1996 are twice as likely as older Republicans to believe that the Earth is warming primarily due to human activities.[9] These are the individuals who are most trusted and therefore most likely to convince other conservatives that climate change is a real issue that affects people's lives and not simply a partisan political label.

Trust-building is equally important in communities where people know one another well but are deeply suspicious of other people's agendas. Garry Peterson, Elena Bennett, and Steve Carpenter at the University of Wisconsin are some of my role models for engaging diverse perspectives around contentious issues. They were concerned about the future of Wisconsin's Northern Highlands Lake District.[10] The lakes of this region attract fishermen and second-home owners from across the state. The resulting population increase brought pollution, overfishing, and a diversity of competing visions about the region's future. Some wanted to restrict development to retain the region's rural character. Others recognized the economic benefits of development. The region had also become a playground for city dwellers from Chicago and elsewhere. Amusement parks, shopping malls, and city glitz were changing the character of some communities. With so many divergent views, community and regional planning bodies had difficulty reaching consensus on plans for the future.

In 2002, Garry and his colleagues organized a workshop that brought together state and federal resource managers and residents with diverse perspectives to discuss potential futures for the region. First there were presentations about the recent history and trends in the region. Workshop participants then divided into small groups and developed 18 scenarios of imaginable futures for the region—not what a particular group wanted but what might potentially happen. Several drivers of future change emerged from the scenario discussions:

- tourism versus new economic opportunities
- external versus local control
- ecological versus non-ecological development
- community ethic versus rugged individualism
- adaptive versus reactive decision-making

Depending on the balance of these forces and other uncertainties, the future of the region might tip in many different directions. This diversity of potential futures encouraged participants to imagine outside-the-box future goals rather than being boxed in by short-term political agendas. Participants chose to focus on flexibility and resilience. This would enable them to be better prepared to learn from and respond proactively to the future as it unfolds.

Based on this project, Steve and his colleagues designed a longer-term scenario project in the Yahara watershed near Madison, Wisconsin.[11] This watershed supports a complex mix of intensive agriculture, rapid urbanization, and recreational and conservation challenges. The project provided a link from *learning*, which was the direct product of the scenarios exercise—as in the Northern Highlands Lake District—to *planning for resilience*. Many of the people who participated in the scenarios were in positions to bring greater flexibility to planning for the region. One woman who was active in designing the scenarios became executive director of a nongovernmental organization that mobilized corporations to help solve the watershed's pollution problems. Participants from the county planning office began considering flooding hazards that they had not considered previously.

The scenarios' exercises provided a politically neutral arena where people could engage respectfully with one another. The combination of diverse perspectives led to new ideas about how to solve problems in both Wisconsin communities. Engagement, listening, and dialogue worked for all these stakeholders, just as they had for conservationists, ranchers, and federal land managers in the Malpai borderlands (Chapter 3).

Sharing Information

The flood of information from promotional mail, television, and the Internet seldom pushes me into action. How can stewardship information be conveyed so compellingly that I and others will act? How can others be convinced to join efforts to address important issues? Simply telling people what we do and why it is important seldom motivates others to join a cause.

I used to think that my role as an ecologist was to discover new knowledge and relay it to other scientists and managers. If my work was potentially useful, I assumed others would read it and that the relevant bits would be passed on to managers and other policymakers who might use it to inform their decisions. I thought of myself as a vehicle for discovery rather than as someone with responsibility to foster change.

How could I have been so naïve? This loading-dock model of communication, in which knowledgeable people produce and deliver knowledge for others to use, is insufficient. This model assumes that the audience has a deficit of important knowledge—like an empty jar waiting to be filled with insight. It assumes that their actions and choices will be empowered by overcoming that knowledge deficit. That's a pretty poor framework for communication. Most managers—and indeed most of the public—know a lot about their local systems and often have strong opinions about what information would be useful. Moreover, managers are too busy to sift through the wheat and chaff of science.

Today's digital revolution has brought an explosion of information delivery through the Internet. Social media now allow anyone to engage in public discussions. What an opportunity! Or is it? We never know all the facts about most issues we want to understand—the environmental impact of alternative products at the store, who to vote for, how best to minimize our impact on the environment. We therefore make the best choices we can within the limits of what we know or believe (**bounded rationality**).[12] The expanded access to information via social media could, in principle, greatly enhance everyone's capacity to make wise choices.

But here's the hitch. Whom do we trust, and how do we know that the information we receive is unbiased? Receiving unbiased news through the Internet is increasingly problematic. Anyone can now post facts, opinions, or lies through blogs and social media. Many people increasingly rely on the Internet as their main source of news. Which of this news is real and which is false?

Search engines now track the interests and beliefs of individual users so that the information and advertisements that appear on each person's screen match what that person previously accessed. In other words, rather than exposing people to a diversity of new perspectives, Facebook and other Internet sites increasingly bombard us with facts and lies that reinforce what we already believe. Perhaps the increasing polarization of today's world is not surprising.

It should be easy for each of us to break out of this vicious cycle, if we choose to do so. However, this requires actively searching the Internet and libraries for multiple information sources about topics of concern rather than passively accepting the opinions that come across our screen. Wikipedia is a good starting point. Both newspapers and independent fact-checking organizations, such as PolitiFact, Snopes, and Skeptical Science, provide independent assessments of claims made by public figures. Ultimately, I suspect that each person must seek out multiple perspectives—especially during election campaigns and other important times. Town-hall meetings and public lectures provide opportunities to engage in face-to-face dialogue about issues that seem clouded in public controversy.

We can't depend on logic and information to shape everyone's behavior, however. One reason that social media have been so effective at reinforcing people's existing biases and beliefs is that many beliefs are socially constructed.[13] Our opinions tend to conform to those of others in our social group (Chapter 6). This explains why gossip and social media play such a powerful role in shaping social dynamics and people's attitudes.

The effectiveness of information and facts in shaping someone else's opinion depends largely on whether that person already has a firm opinion about the topic.[14] If people don't feel strongly about a topic but are interested, their opinion may shift in response to **nudges**—efforts such as information campaigns and hazard warnings that are intended to alter people's behavior in a predictable way without explicit requirements or economic incentives. Nudges such as climate-change information campaigns and warnings of the health risks of smoking are intended to promote social welfare on issues such as health and the environment.

However, people who already have strong opinions about an issue are unlikely to be swayed by information campaigns or hazard warnings. In fact, they are more likely to do the opposite of the behavior that nudges are intended to promote—perhaps increasing climate denial rather than promoting climate action.[15] These people are likely to respond more positively to messengers that they trust than to public information campaigns.

Sometimes information delivery is intended to confuse rather than inform people. A few high-profile scientists have built a career around deliberately discrediting science.[16] They have been funded by a succession of industries that have caused major public-health and environmental problems. The same scientists have been spokespersons who promoted the toxic pesticides used in the early days of industrial agriculture, chlorofluorocarbons that created the ozone hole, the tobacco industry that killed smokers by lung cancer, and the coal and oil industries that are the major causes of acid rain and climate change. These high-profile scientists also receive funding from organizations and industries committed to reducing governmental regulations of public-health and environmental hazards. They have never studied the topics that they speak about and therefore have no firsthand knowledge themselves. Instead, they are "merchants of doubt." They are experts in communication tactics that deliberately seed public doubt and confusion about the validity of scientific consensus.[17] In 1998, climate skeptics circulated a petition claiming that over 30,000 climate scientists believed that climate change was a hoax. However, only 12% of signers had degrees of any kind in earth, environmental, or atmospheric science; and no information was collected to assure the authenticity of the signatures. The actual hoax was the claim by the petition that climate scientists were skeptical of global warming—not that global warming was a hoax.[18] These disinformation campaigns are particularly effective at reaching people who already subscribe to an underlying political philosophy of reducing government regulations.

Although these public disinformation campaigns have been highly influential, I doubt that they completely explain public skepticism about climate change. For a quarter-century, climate scientists have fervently attempted to communicate the causes and serious consequences of climate change.

Surprisingly, as the evidence for climate change and its human causes accumulates, the US public has become more skeptical of climate change and its predominant human origin. About 98% of climate scientists agree that human activities are the major cause of climate change. However, some people pay more attention to the two climate scientists who are skeptical of climate change than to the 98 who are convinced of its human origin. (In a more personal context, would you follow the advice of two doctors who recommend ignoring your child's serious medical symptoms, when 98 other specialists urgently recommend treatment?) Has the delivery of compelling science reduced rather than recruited public support for climate action?

Perhaps climate-change scientists like me have unintentionally contributed to society's disinterest in supporting climate action.[19] The dominant message of climate-change science has been that burning fossil fuels largely accounts for recent climate warming. Emissions have increased because of increases in human population and per capita consumption. The unintended message that comes across is that each of us has personally contributed to climate change by having children, burning fossil fuels, and buying products manufactured from fossil fuels. This negative messaging summons up a host of negative emotions, including fear, anger, guilt, and shame. These emotions trigger deep evolutionary fight-or-flight responses.[20] When these emotions are evoked, people tend to disengage from the problem rather than coming together to solve it.

There is a spectrum of opinions about climate change in the United States.[21] In 2018, about 60% of Americans were deeply concerned about climate change. Another 18% were doubtful or dismissive, and another 22% were uncertain or didn't care. For those people who are concerned about climate change or have not yet decided what to believe, a shift in conversations—from blame about problems to opportunities for solutions—would be more likely to engage them in actions that reduce rates of climate change. Those who are deeply doubtful or dismissive about climate change may only be open to conversations with friends whom they trust.

Searching for Solutions Rather than Blame

In 2014, representatives of Alaskan and Canadian indigenous villages on the Yukon River met in Fairbanks to resolve the issue that had pitted them against one another in bitter controversy for a decade or more. They fundamentally disagreed about the cause of the 15-year decline in chinook (king) salmon, the most culturally and economically valuable fish in the river. Upriver villages blamed downriver villages for taking too many fish and selling them commercially rather than taking only enough for their own use. Downriver villages blamed upriver villages for using highly efficient fishwheels that prevented enough fish from reaching the spawning grounds. Canadian First Nations groups blamed Alaskans for taking so many fish that the United States often failed to meet its escapement-treaty obligations to Canada. Communities on both sides of the border blamed the marine pollock fishery for taking too many salmon as **bycatch**—non-target fish that are required by law to be thrown overboard. Scientific data on chinook were too sparse to support or refute any of these accusations. The only point of agreement was a shared concern that a continued decline in chinook would endanger both the salmon and the cultures of peoples who depend on them.

The meeting in Fairbanks began with a general discussion of the salmon issue and testimonials from villagers along the river about the importance of salmon to their culture and livelihoods. Then a Canadian elder spoke. "We are salmon people," she said. "We need salmon every year—for our family, our guests, and to share with our community. But most chinook have gone away, and we are worried that they will disappear and that our grandchildren will never know what it means to be salmon people. We have not fished for chinook for 10 years. Instead, we fly to Juneau and buy expensive fish. It is the only way we can still be salmon people and save the fish. I hope that someday there will be enough chinook in our river to fish again." The elder's 18-year-old grandson continued this theme, "I have never fished for chinook," he said. "But I know we are salmon people. I've been to fish camp, but we don't fish there anymore because we must protect the fish." The room was silent as the young man sat down.

The next day, the village representatives at the meeting agreed that a moratorium on chinook fishing was essential. Shortly afterward, upriver and downriver villages formed the Yukon River Inter-Tribal Fish Commission and agreed to a moratorium on chinook salmon fishing for that year.[22] Although some people still fished, most did not. This change in conversation from blame to a strategy for solutions hinged on the story told by the elder and her grandson. Their shared concern for salmon as a cultural foundation was more important than next year's fish harvest, and the story suggested options for solutions. Once that concern was voiced and a solution identified, Yukon River villages in both the United States and Canada came together for unified action. In the United States, this consensus opened space for discussions between Yukon River villages and the Alaska Department of Fish and Game about new ways to collaborate in managing the chinook fishery. And in 2017, enough chinook migrated up the Yukon River that the Canadian elder and her grandson fished for chinook for the first time in a decade.

This account illustrates several ways to move difficult conversations beyond the political gridlock that often prevents solutions to contentious issues. Conversations that focus on *positive solutions* and build on care, empathy, and social relationships engage people to come together to solve problems.[23] A positive focus contrasts with *problem-focused* conversations that blame other people for a problem and arouse a host of negative emotions. Respectful conversations with apparent adversaries that emphasize points of agreement rather than disagreement are more likely to open space for new ideas about solutions. Finally, including in conversations people who are most affected by decisions is more likely to foster positive outcomes than are top-down decisions made by people who may be less aware of local contexts.[24] Local people who engage in decisions that affect their lives are also more likely to buy into and support these decisions.

How do we engage in constructive conversation with people who vehemently disagree with us? What if communication were a *dialogue* between people who know the local context and others who recognize new opportunities for improvements? Maybe individuals and groups would then make more progress in communicating information that is relevant, timely, and useful.[25] In practice, people gain knowledge through a wide range of experiences. Sharing different types of knowledge through dialogue provides a more complete picture for informing decisions than expecting people to rely only on scientific or technical experts.

Martha Shulski, Nebraska's state climatologist, is a master of dialogue. She knows how to make people listen when she talks about changing weather patterns. Part of her job requires rigorous quantitative analysis of weather patterns, using complex numerical models and statistics. She knows the weather facts and statistics as well as anyone in the state. The other part of her job is to interact with Nebraska farm groups, community leaders, and politicians to discuss current and possible future weather patterns. They want to know how future weather might affect farmers and communities in the region. She always begins these dialogues with stories. She tells how a recent drought affected the crops of a specific farmer or the quandary faced by a water manager who needed to set water allocations among farms and towns. She only adds the weather facts and statistics if questions arise through dialogue. She uses her weather knowledge to choose representative stories—which tell, in human terms, the impact of well-documented weather patterns.

I've only recently come to appreciate the power of stories in setting the stage for effective dialogue about societal and environmental issues.[26] I should have made this connection long ago. After all, most good news articles begin with a human-interest story. My grandchildren always ask me to repeat their favorite stories, but they've never asked me to repeat any facts I've told them. Scientists like me have been trained to avoid anecdotes (that is, stories) and to base our conclusions only on statistically demonstrable facts. That works well in conversations with other scientists, but it won't work in conversations with more general audiences. The challenge is to tell stories that are consistent with these facts and are also compelling to the audience we want to reach. Journalists have learned this art, so I'm optimistic that scientists and others who want to convey a stewardship message can follow their lead.

Defusing Contentious Conversations

Polarization often reflects a clash of contrasting **worldviews**—that is, ways that people think the world works.[27] How can we broaden public support for stewardship in an increasingly polarized world? Preaching to the choir—discussing important issues only with people whose opinions we share—is comfortable and allows like-minded people to come together to plan outcomes within their reach. That opportunity for collaboration is a useful and important path forward. However, it's mainly an organizational rather than a political challenge. It doesn't open new space for discussions of social actions that might span a broad political spectrum.

Broad public support is important if we hope to shift from today's path of environmental degradation to one of enhancing sustainable conditions for a more habitable planet. Deeper progress will require meaningful and respectful dialogue with people whose actions—consciously or unconsciously—contribute to the degradation of our planet in ways that may differ from our own (conscious or unconscious) impacts. Whether a person intends confrontational or collaborative approaches to problem-solving, understanding other people's motivations and perspectives is important.[28] It may allow agreement on some issues even among people with radically different worldviews—**incompletely theorized agreement** in social-science jargon.[29]

Katharine Hayhoe is my role model for effective communication across deep divides in worldviews. She is a veteran of difficult conversations about how to promote actions on climate change.[30] As a climate scientist, Katharine knows the scientific basis of climate warming and its human causes and consequences. As a communication expert, she is well aware that debates about climate action align largely with political viewpoints, from liberal to conservative. It is a debate fueled by worldviews, not by scientific facts.

Katharine begins every difficult conversation by finding and connecting around shared values.[31] As a mother, she cares deeply about the future her son will inherit. As a Christian, she believes that we all have a responsibility to care for the Earth that God entrusted to us and for the poor, who are disproportionately affected by climate change. She cares about the political and economic security of the United States and the world because these are threatened by climate change. She realizes that accessible energy is important to everyone's life in today's society and understands how renewable energy can meet many of these needs. Her conclusion as a climate scientist is that the most important thing each of us can do to reduce climate change is to talk about it within a framework that we share with others—our family, our neighbors, our friends, our social networks.[32] Seventy-five percent of Americans *never* talk about climate change with *anyone* over the course of a year. Like Katharine, all of us can connect with others around one or more of these issues, regardless of worldviews. Each of these climate-related issues connects more strongly with the heart than the mind, which makes any engaged person an expert who can share their concerns and stories with others. These conversations can build on the care, compassion, and empathy that link people in their daily lives.

Martha's and Katharine's approaches to communication align closely with the 15 principles of persuasive writing distilled by Trish Hall, former op-ed editor of the *New York Times*.[33] They are lessons useful to all of us as we seek to persuade others of our point of view.

I often ask Alaskans whom I don't know well whether they think that the weather has changed since they were young. They usually share unforgettable memories of long stretches of –40°F weather, when the car wouldn't start but the ice was thick enough to drive snowmobiles on carefree outings over lakes and rivers. They also remember being transformed by parents into penguins so bundled in extra layers that they could scarcely walk. All this has changed, they say. Now the thermometer rarely drops to –40° in winter, river ice is unsafe, and summer skies are often choked with a brown smog-like blanket of wildfire smoke that drives people indoors during the months when they most want to be outside. When I cough in October, I still taste last summer's wildfire smoke. As Alaskans, we feel to the depth of our lungs how the weather has changed. There is no controversy here—we share this experience.

If I ask the same people about *climate change*, they sometimes deny its existence. "Climate change isn't real," they may tell me, "it's a hoax." "Climate change" has become a politically charged phrase that is associated with liberal agendas and big government, not with the conservative political views that predominate in Alaska. As Katharine Hayhoe points out, climate change has become a political rather than a scientific issue in the minds of many of these same people who shared their stories of changes brought by new weather patterns.

Because of the importance of worldviews in shaping thoughts and actions, my friends in the Evangelical Environmental Network tell me that "climate change" is not a useful framework to engage their membership. However, environmental changes that harm vulnerable people *are* powerful motivators for their engagement.[34] The facts and solution options are fundamentally similar between "climate change" and "climate vulnerability," but these different framings appeal to people with different worldviews. Engaging in meaningful conversations across a spectrum of worldviews requires careful framing of discussions in ways that respect the worldviews of the audience.[35]

First, though, I must listen respectfully to those worldviews so that I better understand how to reframe the conversation.

When I was a new graduate student in Fairbanks, our family lived on a street named Dead-End Alley. As the name suggests, this dirt/mud track of shacks and outhouses was home to a mix of people who shared a need for cheap housing: other graduate students, miners who came to town for the winter, a resource manager whose boyfriend was a summertime cowboy, a homestead couple expecting their second child, an adjunct English instructor whose husband had fallen on hard times. We came to know one another—and sometimes became close friends—through idle conversations, helping start a neighbor's car, sharing enough coffee to get through breakfast, or helping a neighbor through childbirth when her husband was out of town.

Through these types of informal connections, I became good friends with active members of groups that often bitterly oppose one another in public debates—groups such as the Alaska Miners Association, Greenpeace, the National Rifle Association, and the Union of Concerned Scientists. Most of these individuals were concerned about the needs and rights of people and nature, although many of them thought about these rights differently than I did. Conversations with these friends about shared concerns happened easily, although we usually avoided or joked about hot-button issues that would be conversation stoppers. I therefore see no problem in talking across political and worldview divides, as long as I take the time to listen respectfully, try to understand the reasons that others think differently than I do, and build trust.

These conversations have significantly altered some of my perspectives. When I first came to Alaska in 1969, I viewed the oil and mining industries as vested interests that were hell-bent on destroying Alaska's ecology. Through conversations with friends in the Alaska Miners Association, I learned that this was not always the case. Many companies—like Fowler Construction Company (Chapter 8)—take pains to minimize environmental impacts for both practical and ethical reasons. Abiding by government regulations and developing good environmental performance records improves their future business opportunities.

Oil and mining companies also learned from difficult conversations with ecologists and conservationists. After the discovery of oil at Prudhoe Bay on Alaska's northern coast in 1969, a consortium of oil companies proposed building the Trans-Alaska Pipeline System (with the notable acronym TAPS) to transport oil to ice-free ports in the south.[36] They optimistically expected to complete the pipeline and begin pumping oil within a year or two. They expressed confidence that they could pump hot oil through permafrost, across ice-choked rivers, and across the tectonically active Alaska Range without difficulty. After all, they had plenty of experience building pipelines in Texas and Oklahoma. A series of lawsuits over environmental and land-rights issues put the pipeline plans on hold for 3–4 years. During this time, the oil companies researched how to build a pipeline in arctic conditions. The resulting pipeline was an impressive engineering feat—much safer than the original design. If the pipeline had been built to its original specifications, I suspect that the oil companies would have confronted a damning progression of environmental disasters. They owe a huge debt of gratitude to the environmental groups that forced them into due diligence for their pipeline design.

Is the glass half-empty or half-full? The upside of serious problems is that there are lots of ways to make things better—and that is certainly true for climate change and other social-ecological problems. Focusing on opportunities for improvement—which is the most societally relevant issue—summons up positive emotions that activate thinking and action. It elicits our creative juices and fosters strategies to develop solutions. There are tremendous opportunities to incorporate positive messaging into communications about climate change and other environmental issues. This shifts society's emphasis toward action for solutions rather than a quagmire of deep despair about the problems.

What Can We Do?

Effective communication is critical to learning and sharing stewardship actions and approaches. Beginning with issues about which we are most passionate and skilled,

- **Communicate in ways that build trust**. Strangers or people with a history of conflict often distrust one another. Overcoming distrust requires respectful listening and willingness to understand other people's perspectives.
- **Learn and share knowledge through dialogue**. Dialogue is more effective than "delivering information" as a way to communicate.
- **Focus on solutions more than problems**. Positive messages that steer conversations toward solutions engage other people's interest and ideas

more than do conversations that blame others for problems or paint a picture of gloom and doom.

- **Reframe contentious conversations by emphasizing shared values and goals.** Allow trust to develop before tackling thornier issues.

The next question is: *Given the speed of climate change and other causes of ecological degradation, how can we move most effectively from knowledge and shared perspectives to actions?*

8
Collaborating for Shared Goals

Collaborating to pursue shared goals, as described in this chapter, magnifies each person's impact and is the third element in my four-tiered stewardship strategy.

Groups as Levers for Change

Steve Sayer is a homegrown philosopher who repaired and maintained vehicles at the University of Alaska in the late 1960s. Steve's small wiry frame seemed a poor match for the badly bent bodies of heavy-duty field vehicles. His philosophy for their repair, borrowed from Archimedes, still motivates me today: "Give me a big enough lever and a place to stand, and I can move the world." Ever since then, I've searched for big levers to magnify my own individual effort. But I also look for small ones because they are easier to find and use.

When our kids began school, Mimi joined other parents in leading after-school activities such as a Russian club and skiing classes for beginners. We also joined other parents whose kids were involved in summer sports, music, and nature programs. Parents who emerged from these groups volunteered to coach soccer, teach a foreign language, or lead group music lessons. The school system did not have funding to support these extracurricular programs, so it depended on collaborative efforts by groups of parents. Working together with like-minded people was a lever that enabled parents to provide opportunities for children that would otherwise have been beyond their reach.

Effective groups also emerge, often spontaneously, in the workplace. In 2007, the US Bureau of Land Management (BLM) requested public review of their Environmental Impact Statement of the National Petroleum Reserve-Alaska (NPR-A). The BLM proposed to open this Maine-sized area of federal lands for oil exploration. At the same time, it was tasked with supporting subsistence hunting and fishing by Native American communities and conserving the region's caribou herds and waterfowl. A group of Alaskan graduate students and faculty with expertise in the ecology, culture, and politics of the region decided to review the report as a group. Based on our reading, we discussed the report's strengths and weaknesses. In the end, our input was more balanced and integrated than any analyses we might each have contributed individually. Group review was a powerful lever for learning about the issues and providing effective public input. The next year, one of those students, Stacy Fritz, was hired to manage cultural opportunities in NPR-A. Her broad familiarity with the complex issues of the region made her the logical candidate for the job—and I still go to her when I want to learn how the BLM is negotiating issues that bridge oil, caribou, and native people in Alaska. Collaboration builds connections for future action.

Groups usually have greater impact than do individuals working alone. Besides, it's more fun. We learn from one another, socialize, make new friends, collaborate, and share responsibilities for tasks. Daily life usually brings like-minded people together, creating the opportunity to learn about and join groups that match our interests and goals. However, there are usually more good causes than anyone has time for. The first step is to commit to doing *something*. The second step is to choose groups that match one's values, passions, and skills and have the capacity to trigger change.[1]

Groups that seek to have a broad impact actively recruit others to help pursue their vision. As described in Chapter 3, Wangari Maathai began her tree-planting effort with a few tree seedlings planted in tin cans in her backyard. Her Green Belt Movement eventually recruited women from all over East Africa. They took environmental degradation into their own hands by planting 51 million trees to reduce soil degradation and provide fruit and fuelwood for local communities. Maathai's broad engagement effort massively magnified her lever for change and became a model emulated in many parts of the world.

When public concern bubbles up or problems seem urgent, it becomes easier to form groups that magnify a person's lever and recruit people who share these concerns: school desegregation in the 1960s, Earth Day in the 1970s, the Women's March in 2017. Today's heightened awareness and concern about environmental degradation and social inequities suggest that now may be the time to mobilize people around stewardship for ecosystems and society.

Caring to Collaborate

Many groups coalesce because their members care about the same things. Caring is in our genes (Chapter 1). It is also an integral component of Iñupiat culture in Wainwright, on Alaska's remote northwest coast of the Chukchi Sea.

Local residents look forward to spring not only for the return of light and warmth but also for the spring whaling season, when village crews hunt bowhead whales, just as they have for thousands of years. A harvested whale is shared among all residents in the community—roughly 50 pounds of food per person. With an animal as large as a whale, sharing harvested food makes practical sense. However, sharing in this community extends to all the food that people harvest—fish, ducks, caribou, and berries.

A web of food-sharing connects the entire community, as well as relatives in other communities. It is part of everyday social life. When a hunter shares food with friends or cousins, the recipients, in turn, share with their neighbors and kin, so the same food is often redistributed several times. Shauna BurnSilver, Gary Kofinas, and Jim Magdanz, who have studied food-sharing in two Alaskan coastal communities, tell me that two-thirds of the food-sharing in these communities goes from households that harvest more food to those that harvest less.[2] In this way, sharing especially benefits more vulnerable households, including single mothers and elderly people who are less able to hunt for themselves. Care for other people is a core value that binds communities together—through social relationships, cultural identity, and ties to nature. Sharing is part of being Iñupiat and gives meaning to people's lives. If anything, poor people in these communities are more prone to share because they personally realize its importance.

Along with Mimi and two other grad students, I was unwittingly swept into a web of community caring in the Iñupiat village of Anaktuvuk Pass on our first trip to northern Alaska in 1969. A road had been bulldozed past the village the previous winter, so trucks could transport equipment from Fairbanks, in central Alaska, to Prudhoe Bay, where oil had just been discovered on Alaska's north coast. Trucks had so many problems navigating this winter road that it was abandoned after the first winter. The University of Alaska had sent us to study how the road might be revegetated and the tundra restored. Unfortunately, the wooden stakes and other supplies that we planned to use to set up study plots hadn't arrived, so we spent a lot of time in the first days drinking tea and talking with kids who came to visit our tents across the runway from the village. Every time a plane landed, we asked the pilot if he had the stakes we needed for our study.

As a young ecology student, I wanted to learn about the cotton-grass plants that dominated the tundra near the village. I diligently dug up plants and separated them into their main parts—leaves, rhizomes (belowground stems), and roots—so that I could see how they grew. I didn't know that the cotton-grass rhizomes that I was carefully collecting and separating were also harvested and cached by arctic lemmings for winter food. Local legend had it that, in the old days, Iñupiat people raided these lemming caches of "mouse nuts" to survive during periods of starvation. I continued harvesting mouse nuts, and we drank more tea and kept asking pilots if our stakes had arrived. After a few days, some of the kids brought three women from the village, who took us to their food cellar dug 15 feet into permafrost. We climbed down the ladder and, with the aid of a flashlight, saw about 20 caribou carcasses that had been stored there after the previous winter's harvest. The women gave us a caribou heart and some other meat. They were worried that, since our "steaks" had not arrived, we were reduced to a diet of mouse nuts and tea! Their concern for the welfare of others swept us into their circle of empathy and care.

This empathy for strangers was just the beginning. During the 2 weeks that our group camped at Anaktuvuk, we were absorbed into community activities. The elders invited us to join them in their July Fourth celebration that included fermented whale meat and other delicacies that we didn't recognize (and which the village kids carefully avoided). The kids asked us to join the village dance. They told us that each group of dancers must dance a story, so we danced about eating mouse nuts and swatting mosquitoes. Making fun of ourselves made it easier for others to laugh and talk with us. The next night, as the sun slid along the northern horizon, I was surprised to see a group of kids walking up to our tents about midnight, with fishing poles in hand. They invited me to join them on a fishing trip for arctic char. Several hours later, when one of the younger kids fell down and began crying, because he was too exhausted to keep up, the others stopped to comfort him, built a fire to let him rest, and then helped him the rest of the way back to the village. The people of Anaktuvuk included us in their circle of care because that is what they did. I've had a soft spot for Anaktuvuk ever since—their empathy was infectious.

Just as in tight-knit communities like Wainwright and Anaktuvuk, most people care individually for their children, friends who have encountered misfortune, and places that are meaningful to them. At larger scales, empathy becomes the pathway that enables people to understand, connect emotionally, and act on behalf of those who suffer because of race, poor health, or inability to work.[3] People may also care about those aspects of the environment (the commons) that are not well protected by public or private ownership. However, these societal protections are never fully implemented and are always vulnerable to compromise with short-term goals such as efficiency, profit, power, or simple disinterest.

In recent decades, there has been a gradual abdication by governments of their responsibility to care for vulnerable people and places as job training, healthcare, welfare, and environmental programs are eroded, often in the name of reducing expenses and taxes and boosting corporate opportunities and profit. This trend suggests an increasingly important need to pressure government to meet the needs of vulnerable people and places, as well as to find alternative ways to care for others when government fails to meet these obligations.[4]

Care and empathy for people or places are key elements of stewardship because they move beyond what *I* or *we* want, which is often selfishly motivated, to what society and nature *need* but fail to get. Many of the stories in this book illustrate the power of stewardship, when people who care about other people or places work together to put their concerns into action.

Moreen Miller showed me the power of caring and personal responsibility. At a recent (2018) meeting in Muskoka, Canada, a group of ecologists and environmentalists were exploring the theme of *restoring our relationship with the natural world*.[5] Most speakers talked about the scientific, ethical, and human dimensions of this theme. I was surprised to see the president of a local construction company listed as one of the speakers on the program. Fowler Construction Company operates some 30 gravel pits and quarries in the area to support its construction work. I had always thought of gravel mining as one of the more blatant threats to ecosystems. Moreen, the company president, showed slides of ways that her company restored the aesthetics and ecology of its quarries as best it could, when each operation was completed.

At the end of her presentation, Moreen gave us a quiz—showing us a series of photographs and asking whether each picture was a natural site or a restored quarry. Most of the time the ecologists and environmentalists in the audience guessed wrong—we couldn't tell the difference. The restored sites looked just as natural as rocky habitats that had never been disturbed. By working with ecologists, Moreen's company and other like-minded companies had restored the diversity and natural functions as well as the aesthetics of these sites. When asked why her company and other industry leaders put this effort into restoration, Moreen answered that "it's the right thing to do, and besides it is not as expensive as you might think." Over time, as environmental regulations tightened, her company's experience in restoration enabled it to write better applications for permits. It knew how to restore quarries and had the evidence to prove it. Many of us, as speakers, had struggled to show practical ways to restore human-nature relationships, but Moreen had shown this clearly, not because it was profitable but because it was the right thing to do.

Collective Action

Given the rapid pace of global environmental degradation, isolated steward-ship efforts by individuals are necessary but insufficient to foster the transfor-mational changes needed to move society toward sustainability. Joint efforts made by a group to achieve a common goal (**collective action**) provides oppor-tunities for coordination to achieve more fundamental change.[6] However, co-ordination requires time and effort, so collective action doesn't happen unless people feel strongly about an issue. People are most likely to work together if they have similar perspectives on the issue, feel that their viewpoint has been unfairly ignored, and/or believe that their engagement will make a difference.[7]

In 1980, Candy Lightner's 13-year-old daughter was killed by a drunk hit-and-run driver in California—one of the 50% of traffic fatalities in the United States that are alcohol-related. Each of these deaths is a personal tragedy for surviving families and friends, but what could a single individual possibly do to prevent this from happening again? Candy Lightner triggered a collaboration to address this issue by founding Mothers Against Drunk Driving (MADD).[8] MADD has been rated by the *Chronicle of Philanthropy* as the most popular charity/nonprofit in America and currently has about 2 million members. Its appeal spans the political spectrum, making MADD extremely effective at advocating legislation, such as raising the legal drinking age in the United States to 21 and instituting tougher penalties for drunk driving. The annual number of drunk-driving deaths in the United States has shrunk by 50% since MADD was founded, despite increased numbers of cars and drivers on the road. The tightening of regulations and penalties for drunk driving, largely in response to advocacy by MADD, contributed to the decline in alcohol-related deaths. This shows the power of bringing people together around an issue about which they are passionate and for which there is no reasonable counterargument.

Crisis sometimes precipitates collaboration and action. As the United States emerged from the Great Depression, there was political indecision about what the country should do about political developments in Europe and Asia. The bombing of Pearl Harbor in 1941 galvanized American public support for a war effort in which soldiers who were sent abroad risked their lives and people who remained at home willingly accepted rationing and a decline in standard of living. The entire US economy was redirected toward the war effort in a matter of months, meeting armament production goals that had been ridiculed as impossible.

When the cause is compelling enough, seemingly impossible goals can be attained. The issue of smoking in public places shifted in the 1980s and 1990s from advocacy for the rights of smokers to advocacy for public health and protection from second-hand smoke. Social norms shifted away from public tolerance of smoking as the health risks became more evident. For example, states successfully sued tobacco companies for the added burden of medical care. The smoking issue is not fundamentally different from the current degradation of the global environment, which erodes the rights of future generations. There is increasing consensus that human-induced climate change causes unacceptable risks for both today's society and future generations. This agreement creates the opportunity to launch a campaign to shift social norms (Chapter 9).[9] Many separate policies to reduce human impacts on climate and the environment have been initiated at local-to-global scales but they need public advocacy to ensure that they are understood, incentivized, implemented, and enforced. How can these stewardship components be integrated to trigger the necessary transformation from degradation to stewardship?

Bridging Among Groups

Every important real-world problem is complex. Most are **wicked problems** whose partial solutions create unintended consequences requiring further attention.[10] Integrated approaches are therefore important. Paul Hawken, in his book *Blessed Unrest*, estimated that there are at least 2 million groups worldwide that address different social and environmental aspects of equity and sustainability.[11] Through comparisons with other social movements, he concluded that today's equity and sustainability groups together constitute the largest social movement in the history of humanity. Yet this movement is so dispersed that it is politically almost invisible. These separate efforts seem *less* than the sum of their parts. Groups often compete with one another for members, time, and funds. More fundamentally, Hawken suggested that dispersed attention to the many different dimensions of sustainability may prevent an integrated strategy from emerging.

Perhaps we can learn from instances where groups do collaborate successfully. Identifying these circumstances would be a starting point for developing a more cohesive stewardship movement. Imagine what could happen if these groups coordinated their efforts as "Citizens for a Sustainable Planet," with a commitment to reduce global poverty and rates of species extinction and climate warming by 50%. If MADD can do it with 2 million *members*, why not Citizens for a Sustainable Planet, which could draw on 2 million *groups*?

An initiative in my hometown showed how collaborations can build on one another. The People's Climate March on September 21, 2014, was the largest climate march in history, with 600,000 participants around the world.[12] It was endorsed by more than 1,500 national and international churches and other organizations. Its goal was to voice public concern about climate change to world leaders meeting at the United Nations Climate Summit. Climate-concerned citizens in Fairbanks felt that Alaska needed to be part of this conversation. They therefore organized a local climate march to highlight the irony that Alaska is both a major contributor to climate change through oil, gas, and coal production and a leading victim as rural communities and northern ecosystems disproportionately suffer the consequences of a warming Arctic.

A few months later, in 2015, five people who had been active in the march sat in a Fairbanks bookshop discussing how the community might build on the climate-march momentum and show solidarity with the ongoing Paris Climate Talks. They decided to host a climate movie followed by a public discussion of local climate issues. Within 2 weeks, there were 12 people around the table discussing a longer-term vision and strategy for the group; and within 3 years, there were 750 members. A year after that (2019), the Fairbanks Borough Assembly passed a resolution to formulate a climate action plan to guide local government planning. Thus, the Fairbanks Climate Action Coalition (FCAC) was born and sprang into action in response to emerging local concerns.[13] Each of the initial organizers was active in other local organizations, none of which had a primary focus on climate change. Rather than start yet another single-issue group, they decided to form a coalition with other local groups to integrate various social and environmental dimensions of climate change. **Bridging organizations**, like the FCAC, bring together groups that might not interact under normal circumstances but can accomplish more and different things by working together.[14] Their dispersed structure enables them to recognize and address the many dimensions of wicked problems.

The FCAC consists of several working groups, including an interfaith group that brings people together from various local churches to learn about, pray for, and work toward environmental justice; a group that educates and mobilizes youth around climate change; a renewable-energy group that explores local implementation of renewable energy; a policy-and-politics group that informs voters about the climate-change positions of local political candidates; a regenerative-economy group that explores a just transition from an oil-based to a more diverse Alaskan economy; and an activist group that campaigns to keep oil and gas in the ground.

The FCAC has attracted many people who wear multiple hats, often as members of other groups with related interests. Members of the interfaith group are active members of their churches, and members of the policy-and-politics group are active in political, tribal, and management organizations. These groups now collaborate readily because their social networks span the overlapping concerns of multiple groups. This path toward joint efforts is typical of many local collaborations among organizations. It encourages partner organizations to do much of the work and claim much of the credit for accomplishing specific tasks related to their missions. Together these groups change the story of what is possible in the face of climate change.

Bridging groups sometimes garner massive public support. The Women's March was organized to coincide with the 2017 inauguration of President Trump to protest his views about women and minority groups.[15] Like the FCAC, the Women's March began with a small group—a Facebook invitation from Teresa Shook in Hawaii to a few of her friends to march on Washington in protest. It became the largest single-day protest in the history of the United States, engaging 3–5 million people in the United States (more than 1% of the US population) and another 2 million people in other countries. Perhaps, with the help of social media, the social/environmental movement identified by Paul Hawken is gaining momentum, cohesion, and visibility.

As with the FCAC and major US protest marches, collaboration among groups at large scales requires identification of overlapping interests and often develops through social networks among individuals. The Biodiversity Funders Group (BFG) is an umbrella organization of US nonprofit funders that support environmental stewardship.[16] The mission of the BFG is "to grow a community of biodiversity grant-makers that pursues complementary and collaborative strategies to identify emerging issues, shares knowledge and strategies, and builds partnerships among their member foundations." At the BFG meeting I attended, the members of each foundation seemed well aware of the interests and niche of other groups and made a point of introducing me to people whose interests aligned closely with my own. The BFG had clearly built trust among its member foundations. It had created a culture of sharing social networks and fostering collaboration rather than competition among groups.

Sometimes bridging organizations enable potential adversaries to work together to develop creative solutions that they would not have considered on their own. Renewable NRG Systems, founded in 1982, was one of the pioneers of wind-energy systems.[17] Wind-energy engineers envisioned an energy future that could meet much of the world's energy needs without impacting climate. However, the wind industry quickly learned that wind turbines killed birds and bats, creating a potential conflict between the development of renewable energy and wildlife conservation.

Jim and Wayne Walker, two early leaders in the wind industry, realized the importance of collaborating with conservation and environmental groups at the beginning rather than embarking on a path of legal wrangling and conflict.[18] They were among the founders of the American Wind Wildlife Institute (AWWI), which includes leading conservation, industry, and citizen groups. The Nature Conservancy, a group that was initially concerned about the conservation impacts of wind-energy development, decided to join the AWWI after realizing that only 3.5% of commercially viable wind resources would be sufficient to meet the US energy goal of 20% of energy coming from wind. By collaborating, wind industry and conservation members of the AWWI could jointly identify locations for wind farms that would be profitable, meet local energy needs, and have minimal wildlife impacts. This story is similar to the collaboration among ranchers, conservationists, and federal land managers organized by groups such as the Sand County Foundation and the Malpai Borderlands Group. In these and many other cases, a few visionary leaders invested the effort to bring groups together to collaborate and build trust.[19]

Other bridging organizations build ties between government, science, and citizens. The Arctic Council is an international organization of the eight Arctic nations (Canada, Denmark [including Greenland and the Faroe Islands], Finland, Iceland, Norway, Russia, Sweden, and the United States) and their indigenous residents.[20] The council addresses Arctic-wide issues like the accumulation of pollutants at high latitudes, the conservation of plants and animals, and capacity-building to enhance resilience and adaptation of remote rural communities faced with rapid social and environmental changes. Through their representatives from national governments and tribes, the Arctic Council seeks to implement strategies that span the entire Arctic.

Similarly, at large subnational scales, 20 landscape conservation cooperatives have been established across the United States to facilitate collaboration among federal, state, and nonprofit entities in large regions.[21] This move for coordination and collaboration attempts to replace a historical US pattern in which each government agency managed lands without regard to the management goals and practices of adjoining agencies and private landowners.

Other bridging groups emerge spontaneously as they seek to meet the potentially conflicting needs of diverse stakeholders. The Blackfoot Community

Conservation Area in southern Montana manages the landscape for both conservation and community livelihoods, including agriculture, forestry, and wilderness recreation.[22] The goal of this partnership between agencies and landowners is to develop an ecologically and economically self-sustaining landscape.

Bridging organizations are important because they move beyond preaching to the choir. They open conversations among people with different perspectives about how to solve shared problems.[23] These interfaces between single-issue groups are often the places that generate the most innovative solutions to complex problems—solutions that no single group would imagine or initiate by itself.

Sometimes, too, a single broad-thinking *individual* can trigger these bridging activities. John Bryant, a retired faculty member from the University of Alaska Fairbanks, is fascinated by the interactions among plants, animal browsers, and wildfire in the boreal forest—topics that concern local hunters, resource managers, and scientists who typically work on either plants *or* animals *or* wildfire.

John's research shows that feeding by animals like moose and snowshoe hares causes boreal shrubs and trees to produce toxins that limit further animal feeding. These plant–animal interactions are so complex that no single research group has the knowledge or expertise to address the entire problem. John triggered collaborations among groups of researchers—botanists with organic chemists, plant physiologists with animal population experts, indigenous hunters and trappers with resource managers, population modelers with field experimentalists. In this way, John contributed substantially to the explosion of interest and understanding of the complex dynamics that underlie the response of boreal forests to climate change.

Patricia Cochran, an Iñupiaq woman from western Alaska and executive director of the Alaska Native Science Commission (Chapter 7), is another bridge-builder who makes things happen. She organized a series of workshops across Alaska that brought together native elders, hunters, and trappers with academic scientists to share understanding that comes from indigenous knowledge and Western science. This knowledge-sharing made communities more receptive to university–community collaborations such as the Alaska Community Partnership for Self-Reliance. Patricia—working as a bridge-builder—also brought together indigenous people from coastal Alaska and South Pacific Islands to share experiences and potential solutions to threats from climate-induced sea-level rise. The Indigenous People's Global Summit on Climate Change that she organized bridged among indigenous leaders throughout the world.

Hal Mooney, my former PhD advisor at Stanford University, is a similar bridge-builder. He has repeatedly brought together scientists and policymakers from many nations to address critical environmental problems that are too complex to be tackled by any single discipline or organization. Each of the many issues he has tackled—human and ecological dimensions of climate change, industrial agriculture and food safety, biodiversity and ecosystem processes—is too important to overlook, so he can always convince international colleagues to join him in these efforts. Each of these efforts generates a stack of "Hal-related tasks" that his colleagues both grudgingly and willingly take on as Hal moves on to tackle the next big issue. Despite the short-term frustration of unexpected new commitments, I love and admire him for triggering the types of big changes that the world needs if we are to shift from a pathway of global degradation to stewardship.

There are many similar bridge-builders throughout the world who collectively have the passion, vision, and determination (grit) to shift the planet from a pathway of degradation toward sustainability. These leaders often create temporary bridging activities that bring people and groups together to address a critical emerging issue. All these examples and many others give me confidence that, collectively, society *can* shape a much more sustainable future for our planet. However, the implementation of great ideas is not always easy—or it would have been done long ago.

Collaborations with Government

Many sustainability decisions involve government, which can be slow to adapt and change. In the early 20th century the US federal government established agencies such as the Forest Service and the BLM to manage natural resources on public lands. The intent was to provide "the greatest good to the largest number of people for the longest time."[24] The agency approach to this goal was to maximize resource harvest within the constraints of long-term sustainability. They relied on in-house technical expertise to design and implement appropriate policies. After World War II, an explosion of economic growth increased the demand for resources from public lands. The resulting increase in resource-use intensity gave rise to a top-down management style that emphasized efficient, cost-effective harvest to maximize production of wood, cattle, and other natural resources.

As society became more urbanized in the 1960s and 1970s, this production-focused management style increasingly conflicted with other values, such as recreation and environmental protection of public lands. In addition, increasing public suspicion of "big government" fueled dissatisfaction with top-down agency management styles.

The role of science in management was also changing. A narrow focus on production often led to unexpected environmental disasters.[25] Extensive single-aged stands that regenerate after large-scale forest harvest, for example, promoted the spread of pathogens. Agencies increasingly needed to draw on a wide range of expertise and perspectives from outside of government to solve these complex problems. Only in this way could they provide a scientific foundation for their mandates of multiple use and sustainability on public lands.

Collaborative governance is the collaboration of private citizens and other groups with government to seek consensus in making public decisions.[26] It extends collaboration among citizens to collaboration with government. It's an alternative to adversarial top-down management styles of making and implementing government policies. It also provides potentially neutral arenas for conversations aimed at breaking down the polarization that often impedes government action. However, this collaboration requires a lot of time and effort by both citizens and agency staff—especially on contentious issues. Collaborative governance therefore doesn't always work and may not even be wanted by managers. However, as knowledge becomes increasingly specialized and decision-making processes become more complex, collaboration becomes more valuable. It taps new knowledge and understanding and increases the likelihood of public support for decisions that are made.

As in the Blackfoot Community Conservation Area, the Malpai Borderlands (Chapter 3), and many other contentious resource management battles, collaborative governance (Figure 8.1) is most likely to emerge when no single group has the power or resources to solve the problem on its own— when multiple groups are essential to implement solutions.[27] Collaboration

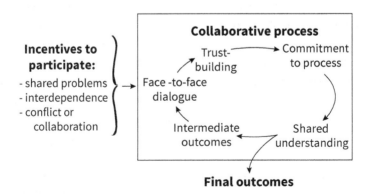

Figure 8.1 Model of collaborative governance.

happens most easily if there is a prior history of cooperation between agencies and other groups. However, even a history of antagonism can lead to collaboration if people recognize that their group can't solve the problem by itself.

The Applegate watershed in southwest Oregon shows how diverse groups of people—fifth-generation farmers, newcomer spiritualists, environmentalists, loggers, and forest and range managers—realized they couldn't solve their problems without working together.[28] As the spotted-owl controversy in the Pacific Northwest forced policy changes on US Forest Service lands, blood pressure rose in the Applegate watershed until opposing groups would talk to one another only in court, and daily life became intolerable. Finally, Jack Shipley, an avid environmentalist, and Jim Neal, a longtime logger, decided that this polarization and gridlock were unacceptable, so they sat down to talk. As their conversation extended to other members of the community, they discovered that everyone wanted a forest that was healthy and intact. The conversation gradually shifted from "us versus them" to "we." Together with representatives of the BLM and the Forest Service and several local environmental groups, Shipley and Neal began to discuss a plan to make the watershed a demonstration site for ecologically and financially responsible resource management.[29] Representatives of various community perspectives elected a board of directors. Nomination to this board required a willingness to work toward solutions and to leave partisanship at home. The successful collaboration that emerged in the Applegate watershed instilled a sense of responsibility for the valley's future. Building trust among different groups through regular communication was critical in the early days of the partnership (as shown in Figure 8.1). Distrust never totally disappeared, but the establishment of a fair and open process prevented distrust from exploding into open conflict.

Collaboration among apparent adversaries requires a process that is fair, inclusive, and transparent and follows rules that build trust among participants.[30] It also requires leadership, perhaps aided by a skilled facilitator, to clarify which values are shared and which are different and to identify opinions that simply reflect misunderstandings. Sometimes people step up as leaders early on—sometimes they emerge from the collaborative process. In general, they are more effective if they are more strongly committed to collaboration aimed at consensus than advocates for a particular outcome. Partial agreement is often possible when the discussion focuses on points of agreement rather than invoking worldviews that divide people into different camps (incompletely theorized agreement).[31]

Compromise is another element that can be important when fundamental disagreements persist. Compromise can emerge in many ways. Regulations might zone different incompatible uses to different places—for example, motorized recreation in one valley and non-motorized recreation in another. In other cases, clarification of opposing values may lead to polarization, political maneuvering, and legal actions that seek to achieve solutions through power and exclusion of one group by another rather than by consensus. The likely success of any strategy depends on local context and leadership. There is no formula to ensure collaboration for stewardship when groups differ strongly in their reasons for valuing a particular place or policy.

In some situations, stewardship collaborations among groups may never develop. Some individuals, institutes, and businesses have a track record of aggressively attempting to thwart any and all stewardship efforts. The Heartland Institute and the Koch brothers, in collaboration with the fossil fuel industry, for example, advocate legislation to promote mining and drilling in protected areas, oppose renewable-energy development, and push to eliminate or weaken environmental regulations.[32] Their opposition to stewardship appears motivated by a mixture of political philosophy (opposition to government regulation and belief that markets can achieve the "best" economic and social outcomes), corporate profit, and consolidation of power. The influence of vested interests is often staggering.

This came home to me starkly at a 2018 workshop held by the Aspen Institute.[33] This international nongovernmental organization fosters discussion of thorny political issues among members of Congress who seek bipartisan solutions to contentious issues. At this workshop, congressional representatives (an equal number of Democrats and Republicans) and subject-matter experts discussed implications of climate change for US energy policy. This was the first time I had met any member of the US Congress, so I was curious to learn their perspectives on climate legislation.

Open discussion of different viewpoints suggested several outcomes that might be broadly acceptable across party lines. Representatives pointed out, however, that many of their colleagues who favored legislation for climate action would never admit this publicly or vote for it because they knew that anti-climate vested interests would find and massively fund candidates to oppose them in the next election. These political pressures by vested interests made fear a stronger motivation than political goals in congressional politics. When money commands, this undermines the effectiveness of democratic processes "of the people, by the people, and for the people" and makes collaboration between citizens, government, and other groups more difficult. The best solution I can see is public exposure of questionable tactics in the media (naming and shaming)[34] and voting for candidates who oppose these tactics.

Interactions with Business

The human evolutionary instinct to compete motivates both people and their businesses to pursue personal gain and profit—among other goals. When the pressure to maximize profit dominates, companies seek to extract as much resource as possible at the least cost. To hell with environmental safeguards! This motivation is often tempered, however, if the person, corporation, or community responsible for resource extraction directly experiences harmful consequences or is motivated by other concerns. Ranchers are unlikely to overgraze the lands they will need for future grazing. Likewise, a community is unlikely to develop or pollute a watershed that provides its only source of clean water, and a company is unlikely to eliminate the supply of raw materials on which it depends.

The profit motive is more problematic when the actions that profit one person or business create costs (externalities) that are borne by the rest of society. An industry that releases pollutants into air or water reduces its own costs of cleaning the waste stream, but society suffers the consequences. A housing development or logged hillside in a scenic area reduces the aesthetic benefits to all, but only the developer profits. Carbon dioxide (CO_2) emitted by one country contributes to the climate warming experienced by all nations, but only the polluting country reaps the short-term economic gains.

In general, lands, waters, and air that are used by all of society (the **commons**) can be sustained only through rules that society agrees to accept. When corporate behavior leaves society with health and environmental burdens, a crisis sometimes serves as a wake-up call to make people realize how much they care about what they have lost. On June 22, 1969, the Cuyahoga River caught fire in downtown Cleveland and became the symbol of out-of-control pollution (Figure 8.2).[35] The river had burned a dozen time in the previous century, so the fire came as no surprise to city residents.

The river was a convenient place to dump industrial wastes. It looked and smelled more like grease and paint than water. Just downstream, where the river entered Lake Erie, algal blooms and dead fish surrounded signs that warned people of pollution's consequences: "No Swimming," "No Boating," "Use at Your Own Risk." Widespread air and water pollution, epitomized by the burning river, motivated the first Earth Day in 1970. Newly elected president Nixon had made air and water pollution top campaign priorities, so there were both grassroots and national pressures to address pollution. Congressional action came through the Clean Air Act of 1970 and the Clean

Figure 8.2 An early (1952) Cuyahoga River fire in downtown Cleveland, similar to the 1969 fire that precipitated the first Earth Day.

Water Act of 1972. I doubt these regulations would have been enacted without widespread public outrage.

Regulations are one way to ensure that polluting industries pay for the resulting damage to nature and human health—the **polluter-pays principle**.[36] This principle is the well-established foundation of most laws that regulate land, air, and water in the United States, Europe, and elsewhere. Publicly elected government sets standards for acceptable levels of pollution, and companies that exceed these standards are fined. Laws and regulations play two important roles in motivating environmental citizenship: they provide guidance about what behavior is acceptable, and they specify the punishment if rules are ignored. If regulations are enforced, most companies comply because they fear the consequences of being caught for disobeying the law. In this way, regulations establish minimum pollution standards and a process for enforcing the behavior that society expects.

CO_2 is now widely recognized by scientists and regulators as *the* major contributor to long-term climate warming (Chapter 4). A simple approach to reducing CO_2 emissions would be to apply the widely accepted polluter-pays principle so that each CO_2 emitter pays for the environmental and social consequences of its emissions.[37] Each CO_2 producer would then have an economic incentive to reduce emissions.

At least two approaches have been suggested as ways to implement the polluter-pays principle for CO_2 emissions[38]:

- **A carbon tax that specifies a *price* (tax) paid to government for each unit of pollution released.** A carbon price of \$40–\$80 per ton of CO_2 emitted would help society meet international climate targets.[39] Most economists favor this approach because they view it as relatively cost-effective and easy to implement—for example, as a tax on coal, gasoline, and other fossil fuels. If designed appropriately, the revenues from such a tax could be distributed equally among all residents of a country. Such a dividend would be economically most advantageous to people with low or modest incomes. This carbon-dividend approach has been advocated by many groups—both liberal[40] and conservative[41]—that are concerned about climate change.

- **A cap-and-trade system that sets a quota (cap) on the *amount* of CO_2 that a region can emit.** This cap can be reduced over time as a way to gradually reduce emissions. Government might, for example, issue 1,000 permits that add up to the regional quota it has decided upon. These permits to emit CO_2 are then either given away or auctioned to companies to define each company's allowable CO_2 emission. Companies pay fines if they emit more than their permits allow. Companies that successfully reduce their emissions below levels specified by their permits can sell additional allowances to other companies that are less successful at reducing emissions. However, the total regional emissions remain within the limits set by government and can be gradually reduced over time.[42] Cap-and-trade regulations played a key role in reducing US emissions of the nitric and sulfuric acids that caused acid rain.

In summary, there are well-established regulatory mechanisms to curtail CO_2 emissions sufficiently to reduce the rate of climate warming, if society chose to do so. Private citizens and their elected representatives can play pivotal roles in discussing how to make these regulations as cost-effective and equitable as possible. The main barrier to progress appears to be clashes among worldviews that prevent genuine dialogue from resolving the underlying issues (Chapter 9).

In some cases, informal rules and community expectations of behavior are more effective than regulations at protecting the commons. Maine has long been famous for its lobsters. Before the 20th century, coastal fishers used small hand-powered boats to set traps for lobsters near their homes.[43] Each family could tend only a few traps, so lobstering had minimal impact on lobster populations. As more people moved to coastal Maine, the lobster trade became more profitable. Gas-powered boats increased the range that a lobster fisher could travel. People set more traps, and competition among lobster fishers became ferocious. Fishers from the same harbor banded together as a "harbor gang" that defended their local waters from outsiders. Outsiders who encroached on a community's territory were likely to be attacked or find their traps cut loose from buoys. People from the same harbor tended to collaborate in protecting their territory because they were part of a tight-knit social network that depended on one another during hard times. Each community made its own informal rules about how many traps each person could set, which prevented local overharvest.

Efforts to pass statewide regulations on catch limits or the number of traps per boat repeatedly failed because differences in environment, lobster density, and access to markets caused optimal fishing rules to differ among communities. In 1995, a law was passed that established seven lobster management zones. Each zone had a democratically selected council with substantial autonomy to set its own trap limits and rules. Today, this local rule-making still guides lobster management.

Green certification and ecolabeling programs are another way to motivate environmentally favorable behavior by business. The Global Ecolabelling Network consists of 27 national and international ecolabeling programs.[44] They certify products and services based on rigorous and standardized health and sustainability criteria. These networks ensure that green labels are accurate and enable customers to choose products and services that minimize environmental impacts. Green certification programs work best for products that consumers care about, such as organic foods and environmentally friendly wood products. The US Green Building Council has established a rating system that certifies buildings for environmentally responsible design, construction, operation, and maintenance.[45]

Green certification is in its infancy but holds great promise. Products are most likely to be certified if a grower or manufacturer wants to advertise the environmental benefits of a product. If consumers seek out green-certified products, businesses are more likely to provide them; and certification networks are more likely to enforce standards across the industry. Expansion of green-certification programs would greatly magnify the role that individual consumers play in fostering environmental sustainability. But it all starts with individuals who commit to purchasing green-labeled products.

In a twist on the green-labeling concept, the environmental and social-justice records of corporations are increasingly accessible, making it easier for investors to choose investments that align with their environmental and societal goals.[46] Many universities, cities, and churches have divested from corporations that mine or use coal and, in some cases, other fossil fuels[47]:

- Universities, such as Johns Hopkins University, King's College London, Stanford University, the University of Maine System, and the University of Massachusetts System
- Cities, such as New York City, Providence, Seattle, and Washington, DC
- Religious groups, such as the Catholic Church, the United Church of Christ, and the Church of England

Sometimes business takes the lead in improving environmental standards. A few transnational corporations (fewer than 10) dominate global markets in most economic sectors. This concentration of economic power in a few corporations characterizes agriculture, forestry, fisheries, energy supply, metals, information, and many other sectors. Many of the top transnational actors have economies that are larger than those of most nations and therefore wield massive influence over human impacts on the environment. This concentration of power in a single company gives it huge potential to promote or undermine stewardship goals.

The World Business Council for Sustainable Development (WBCSD) includes about 200 transnational corporations that together account for 10% of the global gross domestic product.[48] The goal of the WBCSD is "to implement a new agenda for business by laying out pathways in a world in which nine billion people can live well, and within the planet's resources, by mid-century."

Unilever is one example of the WBCSD's corporate strategy.[49] Unilever monitors and reports the environmental impacts of both its own activities and those of its suppliers. It found that local transport of products in diesel trucks and purchase of palm oil from companies that had cleared wetlands for new palm-oil plantations were two of its largest climate impacts. By switching to an electric-powered vehicle fleet for city transport and purchasing palm oil only from companies that produced it sustainably, it was able to reduce both its own carbon footprint and that of its suppliers. A corporation that manages its entire supply chain can estimate its total environmental impact and use this information to influence consumer choice through green labeling.

Groups such as the WBCSD and the Global Ecolabelling Network don't guarantee that business practices become more sustainable, but they provide pathways to encourage it, often in ways that are economically profitable and can be reinforced by the decisions of individual consumers.

If the public is passive about the impacts of industry on the environment and society, profit-seeking motivates corporations to cut as many jobs and environmental safeguards as possible. If citizens join together to advocate for the rights of workers and the environment and to publicly shame businesses with bad environmental and social records, corporations are more likely to be good environmental citizens. Both individual consumer choices and government regulations can incentivize this behavior.

Exploring the Options

Alan Mark is one of New Zealand's foremost ecologists and a veteran of his country's most heated conservation controversies.[50] I visited Alan in 2009 because I wanted to learn how conservation plays out in the messy world of economics, passions, and politics.

Alan told me about the history of Fiordland, a coastal fiord that was on the brink of unrecoverable degradation but today is perhaps the crown jewel of New Zealand's many spectacular landscapes.[51] Fiordland was one of the last areas of New Zealand to be exploited for marine mammals and fish because of its remoteness from commercial ports and roads. Between 1800 and the 1950s, however, the region was successively overexploited for fur seals, right whales, blue cod, and rock lobster. In the late 1950s, road access initiated a recreational fishery with the potential to conflict with the sustainable management of the remaining stocks. In addition, terrorism and armed conflicts in the Mediterranean shifted cruise-ship destinations to safer places, like New Zealand, raising concerns about crowding and pollution.

By the mid-1990s, accelerating resource use in Fiordland by all user groups threatened to destroy the resources that they all valued. In 1995, representatives of all stakeholder groups joined to form the Fiordland Guardians to address the sustainability of the region. After 8 years of facilitated discussion and wrangling, they drafted legislation that formally established the Fiordland Marine Guardians, which was directed to create catch limits for recreational fishers within the fiords and to promote the integrated sustainable management of the Fiordland marine area. The management strategy hammered out in Fiordland was approved by the government and has since been applied to some other coastal areas of New Zealand.

The Fiordland Marine Guardians demonstrate that a shift from environmental deterioration toward sustainability can arise through shared local concerns and initiatives despite diverse and divisive visions for the future of the region. As one of the Fiordland Guardians, Alan knew firsthand that the development of a resource-management plan for Fiordland wasn't easy and depended on the cultivation of trust and commitment to compromise. Each stakeholder group agreed to surrender certain rights in the overall interests of the Fiordland marine environment. Fiordland was the winner—not any of the user groups that originally came to the table. The Fiordland Guardians turned the tide of what seemed to be inevitable deterioration.

What Can We Do?

Collaboration greatly magnifies the contribution that a single individual can make. Individuals can

- **Collaborate with like-minded people to accomplish shared goals.** Seek out groups that share our most important stewardship concerns.
- **Participate actively in the stewardship groups we have chosen.**
- **Bridge with other organizations to broaden our group's impact.** Use and expand our social networks to forge collaborations with other compatible groups around issues of mutual concern.
- **Collaborate with government, business, or adversarial citizen groups, when joint action is essential to solve critical problems.** Conversations with adversarial groups, sometimes with the help of shared acquaintances, can help groups move beyond gridlock.

The next question is: *How do we engage segments of society that do not seem to be motivated by environmental or societal concerns?*

9

Strengthening Democracy

Stewardship emerges most readily from discussions and collaborations built on mutual respect and trust (Chapter 8). However, the world doesn't always work that way. Misunderstanding, distrust, fear, greed, and competing goals fuel many of the interactions that make the world go around. An effective stewardship strategy must therefore be politically astute and address political realities that are sometimes contentious. Political action to foster social and political change, as described in this chapter, is the fourth and final element of my four-tiered stewardship strategy.

Working from Within

Oil has dominated the economy and politics of Alaska since 1969, the year of the first oil discovery at Prudhoe Bay and the year I first came to Alaska. There has been a tug-of-war between conflicting environmental claims by industry and environmental groups, with science often caught somewhere in the middle. The underlying politics have often been nasty, making it difficult for citizens to evaluate the self-righteous claims that each group has made in public media.

Bill Streever is a hard-nosed pragmatist who directed environmental and ecological research for BP (formerly British Petroleum) for 16 years in Alaska. He wanted to know how oilfield operations and associated infrastructure at Prudhoe Bay affected the local ecology and what could be done to minimize undesirable impacts. For example, what were the effects of nest predation by ravens on the nesting success of threatened species of waterfowl, and how could predation be reduced by controlling raven access to garbage and potential nest sites on oilfield infrastructure? Many of these BP-supported studies addressed potentially contentious environmental issues, but Bill focused on informing and developing stewardship solutions rather than engaging in politics or **greenwashing**—unsubstantiated claims of the environmental benefits of a practice or product. He was careful to ensure that studies were done objectively and rigorously and that data were shared with non-company scientists, often through peer-reviewed publications. The long-term ecological monitoring program he initiated within BP provides baseline data that will help detect problems that emerge. Documentation of ecological changes in more than 100 tundra rehabilitation sites in the Prudhoe Bay oilfield provides guidance for future restoration as aging parts of the oilfields move toward abandonment.

By *working within* BP, Bill did a lot to restore and maintain the credibility of oil-company science in Alaska, which in turn led to a more collaborative working relationship with non-company scientists. He was an effective advocate for environmental stewardship within his company and changed my opinion about important roles that environmental stewards can play within industry.

ABR (formerly Alaska Biological Research) is an Alaska-based environmental consulting company that emphasizes the triple bottom line of economic viability, environmental sustainability, and social responsibility.[1] Over the years, ABR has hired some of the most talented and environmentally concerned ecology graduates from the University of Alaska. ABR practices what it preaches in terms of minimizing negative environmental and social impacts of its operations. It often accepts contracts with industry and government agencies to help find solutions to unintended ecological disasters. ABR focuses on fixing the problems rather than blaming the actors. It is widely respected by both environmental groups and industry for its management style and focus on solutions.

Many people who work in government agencies are also passionate environmental stewards. My first glimpse of the inner workings of federal agencies came as an author and advisory committee member of the 2014 US National Climate Assessment (NCA).[2] Every national and international climate assessment over the previous 15 years had been accused by climate-change skeptics of misrepresenting the facts about climate change. And, indeed, it is important to identify and correct any errors, no matter how trivial, that inevitably creep into complex environmental assessments because any mistake or overstatement weakens the credibility of the entire effort.

The review process during the 4-year writing of the NCA was mind-numbingly thorough. Every important NCA decision had to be approved by consensus of the 60-person advisory committee, which often meant lengthy debates about wording or details of the NCA process. Each scientific statement in each chapter had to be backed up by published studies. Conclusions required detailed explanation whenever there was conflicting evidence.

The reviews—internal reviews, external reviews, checks for consistency of statements among chapters—went on endlessly. Every comment and question raised by any reviewer required a response from the author team, which in turn was reviewed by a review editor to ensure the validity of the response. The reviews that were most critical of the science received the most careful attention. In this way, climate skeptics had a direct (and disproportionately large) opportunity to provide input to the balance of evidence reported in the assessment. I was impressed that, when a federal agency is committed to a task, it can do its job with incredible thoroughness. The result, in the case of the NCA, was an assessment of climate change and its impacts in the United States that was highly credible, although somewhat conservative, given the need to document evidence for every statement in the report. I was impressed with the conscientiousness of all the people who worked on this government assessment, either as government employees or as volunteers from outside government. I had a similar positive experience participating in the Intergovernmental Panel on Climate Change assessment,[3] the international counterpart of the NCA.

However, the workings of government are not always a smooth ride. In March 2001, Brad Griffith was on a plane from Alaska to Washington to testify before Congress. As a caribou biologist in the US Geological Survey, he knew that he was headed for the hot seat. The Arctic National Wildlife Refuge (ANWR) had become a line in the sand drawn, on one side, by President George W. Bush, who was determined to open the ANWR to oil development and, on the other side, by national conservation groups that were equally committed to keeping development out of the refuge.[4] At stake was a portion of the ANWR's coastal plain, which was underlain by significant oil- and gas-bearing formations about 150 miles east of Prudhoe Bay, Alaska's most productive oilfield. This area was also the calving grounds for the Porcupine caribou herd, one of the main reasons for the initial establishment of the ANWR as an area to be protected from development. Brad was the caribou biologist who led the research on calving patterns of the Porcupine caribou herd. He had also participated in assessments of the impacts of oil development on caribou at Prudhoe Bay. Based on this experience, Brad had developed models of the likely impact of different levels and locations of oil development on caribou-calving success and population size. He knew that his research findings would be controversial, no matter what they showed.

Brad has a track record of research on controversial issues, in part because he is drawn to the excitement of making science relevant to tough management issues. In his words, "It would be boring to do research that has no policy relevance. Why would I want to do that? I want my research to make a difference." He knew that his science about caribou in the ANWR had to be bulletproof. It was certain to be attacked by parties on both sides of the ANWR's development debate. Brad concluded that oil development at the western edge of the ANWR's coastal plain would probably have no detectable effect on caribou populations but that more extensive development would probably curtail calving success enough to reduce caribou populations. He refused to be drawn into statements or interpretations that supported either pro- or anti-development positions rather than science. According to Brad, he later learned that high-level administrators had ordered his boss to fire him for not falling into line with the government's official position. However, his boss refused to fire him because he knew Brad's science was solid. As of spring 2020 the ANWR coastal plain has not been developed. Brad made a difference by working within his government agency—not by being politically safe or opportunistic but by sticking to his principles of rigorous science.

Deepening Democracy

Pavel Borodin, mayor of Yakutsk, a large Siberian city, leaned across the table to my wife Mimi and whispered, "Who is the most powerful person here?" She looked around the large hotel banquet room filled with local Fairbanks and Alaska-level government and industry leaders gathered to hear a presentation about oil and gas exploration. Knowing who was most powerful was difficult because all the dignitaries played such different roles.

The year was 1989, just before the breakup of the Soviet Union. The cities of Fairbanks and Yakutsk were exploring the possibility of becoming sister cities, a natural fit because of their similarities in environment and resource-extraction economies. Pavel clearly wanted to understand the power dynamics of our country. He pushed a bit harder, "Is the mayor the most powerful person?"

"No, not at all. She serves at the pleasure of the people," Mimi answered.

"Yes, of course," Pavel snorted, giving a dismissive wave of his hand. "We say the same thing." He clearly had no idea how hotly contested and close American elections can be nor how much scrutiny American mayors can receive.

Many years later, the sister-city relationship had blossomed; and Karl Kassel, then director of the Fairbanks Parks and Recreation Department, was hosting another delegation of Russians from Yakutsk. As they toured the Fairbanks North Star Borough office building, the delegation had many questions about the Fairbanks budget. Karl said, "Let me show you," and he pulled up the current line-item budget for the borough. The visitors were amazed that he could access it so easily and was willing to show it so openly. Karl responded, "We have to keep the budget at our fingertips. Any citizen has the right to see it at any time."

After many citizen and scientific exchange visits to Russia, Mimi and I feel lucky to live in a democracy, despite daily reminders of its warts and flaws. The accounts of political corruption, vested interests, and power politics in the United States often seem similar to those in Russia. However, citizens in a democracy have the opportunity and responsibility to select their leaders in ways that can shape the future of their community and nation as well as constrain the nastiness of politics.

If most people in a democracy prioritize a sustainable future of their community or nation, they can elect leaders who pursue that vision. More often than not, however, election debates degenerate into a focus on short-term concerns or generic worldviews, rather than on a thoughtful discussion of ways to shape the future. It doesn't need to be that way, if people reframe the debates and actively engage in the political process. When politics prevents this from happening, democracy is not living up to its potential.

Democracy has always been a work in progress. The framers of the US Constitution debated ferociously about how "we the people" should govern, as did the citizens of Athens 2,000 years earlier. Many of the debates at the time of the American Revolution centered on who qualifies as a citizen and how citizens' votes translate into selection of leaders. Access to citizenship and the opportunity to vote in the United States remain controversial as people emigrate from other countries and as powerful people tinker with rules to either favor their own voter base or broaden opportunities for public participation.

US democracy has changed a lot since I first voted in 1965. Social movements surrounding civil rights, the environment, and women's issues have galvanized the public and raised legal requirements for public consultation in policy decisions.[5] Meanwhile, government has generally retreated from its role in supporting public well-being—instead privatizing components of health and public services and leaving the public to fend for itself, often without public resources to do so. Thus, *we the people* are caught in the middle, with greater opportunity and necessity to participate in the decisions that shape our lives.

The League of Women Voters (LWV) was founded in the United States in 1920, in preparation for the 19th Amendment, which gave women the right to vote.[6] The LWV began as a massive political experiment to help 20 million women carry out their new responsibility as voters. Its vision has now expanded to fostering conditions in which every person can have the desire, right, knowledge, and confidence to participate actively in democracy. The LWV is explicitly non-partisan and focuses on deepening the democratic process without telling people what they should believe. It focuses on clarifying election issues, informing voters about their choices, and improving opportunities to vote.

The LWV is not alone. The American Civil Liberties Union and the Southern Poverty Law Center speak out for vulnerable people, and The Nature Conservancy and the Ecological Society of America speak out for ecosystems. These are just a few of the many organizations and voices concerned about the future of our planet.

There are many groups of like-minded individuals who coalesce around particular issues. They craft statements that articulate their vision for the future and the steps needed to get there. The Fairbanks Climate Action Coalition, to choose a local Alaskan example (Chapter 8), developed such a statement as a basis for discussions with legislators, letters to the newspaper, town meetings to spread the word, and protests against political actions that undermine progress toward climate solutions.[7] Vested interests will, of course, also seek to influence political outcomes through lobbying, campaign donations, and political favors. However, a group with broad public appeal has the advantage of bringing more votes to the ballot box. At times like the present, when politics often seem dominated by powerful interests that can buy or trade political favors to benefit themselves, it is all the more important that people who care passionately about a healthy and just future for the planet speak loudly and vote conscientiously in support of their beliefs.

A successful ballot initiative or candidate must appeal to a majority of voters. If a group's core issue lacks the necessary breadth of public support, the issue must be modified or reframed to broaden its appeal. This usually requires respectful engagement with other groups that might be concerned about the issue from different perspectives. Some members of the Fairbanks Climate Action Coalition participate in an interfaith prayer group that seeks to raise public consciousness about environmental- and social-justice issues. In these broader discussions, the climate-action agenda often shares the stage with other issues of mutual concern, such as the plight of vulnerable people or care for Creation. With time and respectful discussion, a constellation of interrelated issues can surface and provide a broader basis for dialogue, consensus, and action on topics such as environmental justice or climate change.

No group can accomplish much without leaders and spokespersons who can articulate the group's concerns. Christina Salmon (Chapter 3) of Igiugig in southwest Alaska was born with the right surname. She cares deeply about the future of salmon. At first glance, she might seem like the least likely candidate to run successfully for the Lake and Peninsula Borough Assembly, a body representing an area the size of West Virginia. After all, the 70 residents in her village don't come close to the number of votes needed to win a borough election. Most of the people in Igiugig care about salmon as a cultural and food resource rather than a source of employment. However, residents of larger towns with commercial fisheries and canneries are more likely to support candidates advocating for rural employment. But they all care about salmon. Moreover, Christina is the most networked person I know—through friends and extended family in other villages and through Facebook. By framing her campaign on the future of salmon and encouraging her network of friends and others concerned about salmon to vote for her, she was elected to the borough assembly, which now actively opposes the development of large mining operations in the region. If implemented, these operations would threaten the water quality and habitat of the largest remaining sockeye salmon run in the world.

Leadership is only one of many ways to deepen the democratic process. In 2007, Alaska governor Sarah Palin asked about 100 climate-change experts from academia, business, government agencies, tribes, and other groups to form the Alaska Sub-Cabinet on Climate Change to advise her on what Alaska should do to address climate change.[8] This was one of the most momentous occasions in my professional career because no one from government had ever asked my opinion about climate change, even though I had researched the topic for 30 years. I had assumed that my role as a scientist was to do research and that someone in Government (big "G") would eventually sift through this stuff and use whatever seemed relevant to design policies. This loading-dock model of delivering scientific information to policymakers didn't seem to work, so imagine my excitement to be asked directly by government for advice. Wow!

I was a member of the team tasked with examining the impact of climate change on Alaska's natural resources. Other teams dealt with health, the energy sector, and climate-related flooding and erosion. Our team worked well together, even though few of us had known one another beforehand. Over the course of 2 years, we came up with a set of prioritized actions that Alaskan agencies could implement to reduce climate-change impacts on Alaska's natural resources. We thought carefully about political realities, economic feasibility, the likelihood of favorable outcomes of each option, and how it might be implemented. By the time we proudly presented our report to the governor's office, however, the political winds had changed. Sarah Palin had resigned the governorship and set her sights on bigger goals, and the new governor said *thank you very much* and put our report on a shelf, where, as near as I can tell, it has sat ever since.

My experience with Governor Palin's subcabinet shows both the opportunities and shortcomings of public engagement in informing, consulting, and advising—the forms of participation that are most widely specified in legislation.[9] Many laws, such as the National Environmental Policy Act require federal agencies to "provide meaningful opportunities for public participation."[10] In other words, the opportunity for public participation is legally required.

I was disappointed that the subcabinet's report seemed to have no impact on Alaska's climate-change policy. The downside of public participation—as informing, consulting, and advising—is that public participants have no designated role in making decisions, and their input may not even be wanted by decision makers.[11] The upside is that participation provides the public with direct access to decision makers, and sometimes this advice is useful to them and makes a difference in the outcome. Although the Alaska governor's office never officially accepted or promoted any of the report's recommendations, individuals within state agencies knew about the report and have implemented most of its recommendations, by working beneath the radar of public attention.

In 2018, 11 years after the initial effort, the Alaska governor created a new task force to design a climate-change strategy for Alaska.[12] The earlier climate-change report served as a starting point for their new planning process. However, once again, the current governor (2020) has shelved the climate-planning process. Now the initiative for climate planning is shifting to cities, tribes, and citizen groups like the Fairbanks Climate Action Coalition. The social networks formed during the previous statewide planning processes now help connect local efforts across the state.

Participation in the consultation process also changed my attitudes. I now know and network more closely with people in tribes, government agencies, and business about climate-change issues; and together we have a better understanding of one another's perspectives and who to contact to learn about issues or to inform others who make decisions. Based on this expanded social network and my more flexible schedule in retirement, I now seek out opportunities to contribute to Alaska's climate-change planning process, rather than sitting back and waiting to be asked.

Others, like Robin Bronen, engage more immediately in the political process. When Robin believes that government is falling short of its responsibilities, she looks for ways that she, as a private citizen, can improve the situation. As an Alaska human-rights lawyer, Robin works with immigrants seeking asylum in the United States. Her work took her to indigenous villages in coastal Alaska, where she saw firsthand that communities often faced life-threatening erosion with each fall storm. However, communities had no legal access to funding to move to safer ground. Robin realized that this dilemma was similar to situations in many coastal regions of the world and that the increasing number of people needing to escape sea-level rise was an emerging international human-rights crisis.

As a lawyer rather than a scientist, Robin didn't have the scientific expertise to contribute to solutions. She therefore returned to school and completed a PhD on the legal barriers to climate-change adaptation in Alaska. She and other students and scholars at the United Nations University articulated the legal rights of people displaced by climate change. Robin then wrote articles in legal and scientific journals about the legal and political barriers that prevent Alaskan coastal communities from relocating to escape life-threatening erosion.[13] She and others also participated in discussions with Alaskan communities and government agencies to find work-around solutions that would enable communities to protect themselves within the current legal framework. By the time Robin had completed her PhD, she was a recognized authority on policies related to climate-induced coastal erosion. In 2015, President Obama invited her to join a group that was drafting climate policy. Alaskan communities have also invited her to collaborate in research that enables them to document the climate threats they are experiencing. Robin now knows many of the pressure points where she, as a private citizen with relevant professional skills, can make a difference.

There are many ways to participate actively in democratic and community processes that foster stewardship. Potential commitments range from a few hours per year to Robin's total immersion. They also range from actions requiring no expertise to those benefiting from sophisticated understanding or personal connections to powerful people: picking up litter during spring cleanup; networking with a powerful politician who is an old family friend; writing letters that comment on local resource-management plans or environmental impact statements; voter registration in underserved communities; donation of time or money to support candidates or causes we believe in; running for office; volunteering in community efforts such as ranchers' associations, stream restoration projects, church-supported soup kitchens, and so forth. All these activities, and many more, foster a more sustainable future for ecosystems and society. I believe that personal commitment to stewardship actions is a universal human responsibility. Today's world is blessed with so many solvable problems that each of us can contribute to solutions that match our personal passions, skills, connections, and levels of commitment. The tragedy would be to abdicate this responsibility and contribute only to actions that degrade our planet.

Reframing Debates

Political gridlock often results from heated debate about **false dichotomies**— complex sets of issues that are described simplistically as a choice between goals that are emblematic of different worldviews—for example, jobs versus the environment or climate action versus the economy. In practice, the underlying issues are usually complex wicked problems that probably have no simple solution (Chapter 8). However, I've seldom seen false dichotomies unpacked to the point that substantive issues can be rationally discussed, trade-offs identified, and solutions designed. However, there are logical steps to move from irrational political gridlock to genuine discussion of solutions.

In the spirit of respectful dialogue, the *first step* is to defuse politically charged debates by avoiding stereotypes that are intended to anger and polarize the audience. Speeches, social media posts, and (dis)information campaigns that paint broad segments of society (such as the spectrum from liberal to conservative) in terms of evocative stereotypes, such as racists or communists or atheists or fascists, invite citizens to act based on fear and anger. People then seek to demonize their "enemies" rather than search for common ground.

The *second step* in reframing debates is to unpack complex social and political problems into substantive issues that require genuine discussion of information and ideas. Neutral parties, such as the LWV or facilitators who have experience in fostering true debate, can often play constructive roles. The Aspen Institute (Chapter 8) is one such neutral party. Each of its workshops invites an equal number of Republican and Democratic members of Congress to avoid domination of debate and discussion by one political faction.

The "jobs versus environment" debate is a good example of a set of issues that is not nearly as simple as is often portrayed. In an effort to remain competitive, businesses often invest in labor-saving innovations such as robotics, online information exchange, and automatic tellers to reduce their labor requirement and costs. Some businesses also buy inexpensive components from overseas or relocate to countries or regions with lower wages. The net effect is fewer jobs, changes in locations of available jobs, and more skills required for many of the remaining jobs. In addition, technological innovation changes markets, so some industries expand and others contract. Today, there are far fewer jobs than in the past for telephone operators, tailors, and shoe-repair people but more jobs for people with computer skills.

Concurrent with these changes in labor markets is improved information about the levels and the environmental and health consequences of pollutants produced during manufacturing. This improved information leads to regulations that protect society from serious unintended consequences. Each of these separate changes in labor markets and regulations is a well-intentioned effort to improve efficiency and promote societal well-being. However, these changes interact and often have unexpected consequences that play out badly for some people in some places. Changes in the coal industry illustrate the complexity that underlies the false dichotomy that is often portrayed simplistically as a choice between jobs and the environment (Box 9.1).[14]

Individuals play important roles in solving complex problems by reframing discussions around real issues and solutions rather than retreating behind generic worldviews that are simply a smokescreen that evokes fear and anger and prevents progress.

Development versus conservation is another false dichotomy that is fueled by clashes among worldviews but hides opportunities for solutions that would simultaneously improve both human well-being and species conservation.[15] There are many ways to sustain or enhance society's capacity to thrive while protecting nature's biodiversity, but it requires genuine debate about real issues and fast action to reduce human impacts on climate (Chapter 4). Examples of other simplistic false dichotomies that are critical to the implementation of stewardship include the following:

- Climate action versus the economy
- Government regulations versus free markets
- Pesticides versus the environment
- Reduced taxes versus citizen welfare
- Trickle-down economic benefits versus bottom-up rights and opportunities of workers
- Individual rights versus social responsibility

These and many other issues are fundamental to stewardship and a vibrant democracy and economy and worthy of meaningful discussion. Research on the underlying dimensions in each of these false dichotomies can lay the groundwork for genuine solutions. These are fixable dilemmas that would benefit from deeper public and professional engagement. The current simplistic arguments fueled by clashes between worldviews don't provide rational

Box 9.1 **50 Years of Change in the US Coal Industry**

Coal production in the United States tripled between 1960 and 2009. During this same time period, labor productivity increased 18-fold—each miner could produce 18 times more coal in 2009 than in 1960. Coal jobs were therefore eliminated despite increased production. In the eastern United States, a new technology (mountaintop removal) began to replace underground coal mining and greatly reduced the number of workers needed to access a given amount of coal. Concurrently, after 2009, the expansion of hydraulic fracturing (fracking) reduced the price of natural gas, so it was cheaper to produce electricity from natural gas than from coal, especially from older coal-fired power plants, which had high maintenance costs. Many of the older coal-fired plants were therefore closed and jobs eliminated not because of environmental concerns but because of innovation by industry.

So how do the environment and government regulations play into this coal story? The Clean Air Act of 1970 restricted sulfur emissions from new coal-fired power plants. Older coal-fired plants, largely in the East, were exempted from this regulation in order to get enough votes to pass the legislation. These older power plants continued to use eastern high-sulfur coal. However, western coal, with its much lower sulfur content, was more suitable for meeting the new emission standards, so most of the expansion of coal mining since 1970 occurred in the West.

Two additional factors amplified the competitive advantage of western coal producers: (1) they had a sixfold higher labor productivity (only one-sixth as many workers needed to produce a given amount of coal) and (2) deregulation of railroads in 1980 caused railroads to reduce their price for transporting coal by 50%, so western coal was cheaper—even in parts of the East. The result was stable low employment in the coal industry in the West (about 20,000 jobs) and loss of about 115,000 jobs (57% of coal jobs) in the East. Of course, the story is more complex than this. However, the bottom line is that, as in many industries, the coal industry modernized faster than it expanded its production, so it needed fewer miners. Both environmental regulations (supported by environmentally concerned citizens) and railroad deregulation (supported by people opposed to government regulations) benefited the coal industry in the West. This plus the closure of older uneconomical coal power plants in the East and the price advantage of natural gas over coal led to further loss of coal jobs in the East. These changes are clearly not a simple story of coal jobs versus the environment. The changes also raised a new set of environmental issues as technology changed—such as the environmental impacts of mountaintop removal and fracking.

But does a deeper understanding of the issues move us toward solutions? Although there isn't a simple answer, examples of topics that might contribute to solutions include the following: What types of job training and economic development programs in eastern coal-mining areas could build on the pride, self-respect, and skills of coal miners as contributors to society? Can innovative research on carbon capture reduce the carbon emissions from coal-fired power plants enough to make coal competitive as a carbon-clean fuel—in other words, to produce more electricity per unit of carbon emission? This is a technological and scientific question—not a political one.

Even though it is unlikely that coal, oil, or natural gas could produce electricity with lower carbon emissions than renewable-energy technologies, all of these energy sources will probably continue to be used in some situations. It therefore makes sense to pursue research that would make each of them as carbon-clean and cost-effective as possible. Each of these issues is worthy of debate and innovative solutions.

pathways to solutions. Instead, the engagement of news reporters, individual citizens, and non-partisan groups like the LWV is required to probe these issues with greater depth and respect.

How can false dichotomies be identified before they explode into hardened public controversy? Any complex issue where change creates winners and losers can mushroom into a false dichotomy, especially if people differ in the attitudes that shape their opinions. For any highly visible public issue, it's worth asking who wins and who loses. It's probably easiest to engage in respectful debate about underlying complexities before an issue explodes into full-blown controversy. In these early stages, respectful conversations are more likely to discover opportunities for solutions (Figure 8.1).

Some actions by businesses or individuals cause unquestionable harm to the environment and create many losers and only a few winners. Although pesticides are intended to reduce crop pests, they spread to adjoining and downstream ecosystems where they kill non-target organisms and often endanger human health. They also impose strong evolutionary selection on target pests, leading to the evolution of pesticide-resistant pests.[16] This explains why US farmers lost almost twice as large a proportion of their crops to pests in the 1980s and 1990s as in the 1940s, despite greater pesticide use. Nonetheless, in any given year, pesticides are essential to some farming practices and provide profits to pesticide manufacturers and their investors, so it's difficult for society to step back and explore novel solution pathways— although these alternatives are readily available.[17]

Other industrial products or activities that are unquestionably damaging to the environment include overuse of antibiotics, excessive fertilizer application in agriculture, tropical deforestation, overfishing, emissions of greenhouse gases, and accumulation of toxic mine wastes. Each of these products and activities generates profits for industry and benefits for some segments of society. Broad discussion involving experts and the public provides an opportunity to identify and evaluate key issues and trade-offs and to explore alternative solution paths. These are among the rich dimensions of democracy that are seldom explored but require input and leadership from private citizens.

Challenging Vested Interests

I grew up in Chapel Hill, North Carolina, in a segregated world—schools, dances, sports, lunch counters, movie theaters, public restrooms, buses, and so forth. Although I believed segregation was wrong, I didn't see what I could do about it. Discussions of the issues in my church youth group raised my awareness but seemed academic. The inspirational leadership of Mahatma Gandhi and Martin Luther King, Jr., didn't seem relevant to an ordinary teenager like me.

In 1960, nine African-American high school students asked to be served at a drugstore lunch counter in Chapel Hill.[18] They refused to leave when asked and were arrested. They were the same age that I was! Segregation suddenly seemed much more personal. I paid more attention to the news about marches and sit-ins.

A couple of years later, several of my friends and I began attending rallies at the local African-American church where we made friends with African-American kids we had never met (Prologue). We learned to internalize an ethic of nonviolence and took our turns on the picket line at the drugstore and movie theater. In 1963, the local Merchants Association that advocated for segregation of businesses became a focus of protests. The policeman who arrested me for participating in the sit-in at the Merchants Association was the same man who had supervised the weekly roller skating on a blocked-off neighborhood side street when I was a kid, so he knew me as part of the community (Figure 9.1).[19] We thought of our time in jail as a new place to express our concerns rather than cruel punishment. We were part of the community, and people listened to us.

Formal desegregation of businesses and schools has occurred in Chapel Hill, although inequities remain. Civic leaders from the African-American and white communities had been meeting for several years to map out a strategy for gradual desegregation. The community as a whole was more liberal than most nearby communities. The sit-ins were mostly by local residents and targeted the most visible and vocal bastions of segregation. City council meetings provided a less politically charged formal venue for dialogue among decision makers who had broader authority than either the demonstrators or

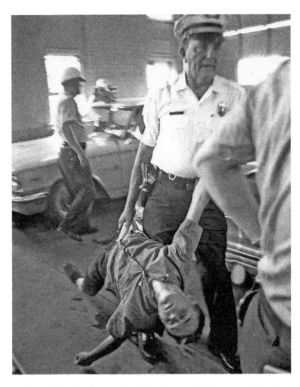

Figure 9.1 Stuart Chapin being hauled to jail after a 1963 sit-in at the Merchants Association offices in Chapel Hill, North Carolina.

the merchants. Perhaps the main impact of the sit-ins was to raise visibility and awareness among civic leaders and the general public about the reality of segregation in a community that prided itself for being progressive.

My next arrest for protesting segregation was in a low-income, strongly segregated community near my college in Pennsylvania.[20] This time (November 1963), the community probably considered me an "outside agitator." The arresting policeman pulled me down the stairs by my feet and threw me in a van, and we protesters were taken to a regional prison and locked up with all the other inmates. To the best of my knowledge, there was no broad public support among white residents in the community to address issues of inequality, so the sit-in probably polarized rather than informed the community. At my recent 50th college reunion, I learned from a classmate who continued to work in that community that the racial disparities in power and opportunities have largely persisted over the last half-century.

Marches, rallies, and other nonviolent protests have contributed to the success of many important social movements but not because they alter the opinions of their adversaries.[21] Instead, successful protests increase the level of engagement of participants and raise the visibility of issues among people who may not have developed strong opinions about the underlying controversies.[22] Protests have the greatest long-lasting effects when they have a clear message, are nonviolent, and are part of well-organized sustained efforts with broad appeal. For example, across 2,800 US counties, those that had peaceful civil rights protests in the 1960s now have less racism and are more likely to support affirmative action 50 years later compared to similar counties without a history of peaceful protest.[23] Protests also influenced all branches of US government: Congress, the presidency, and the Supreme Court.[24] Successful protests and social movements can lead to transformation.

In 1999, a group of about 15 international scholars teamed up as the Resilience Alliance.[25] They realized that many of the world's most serious problems require a fundamental transformation of society's relationship with nature.[26] I joined this group in 2004 because I was concerned about the same issues. We often met in Stockholm, which was why I chose to go there in 2007 to write my book about stewardship (Chapter 1).

The Resilience Alliance had studied the transformation of many systems around the world[27]—including the ones described in the next section. Their studies suggested several strategy elements that fostered the planning and implementation of successful transformations and that built resilience of the systems that emerged when transformation occurred (Box 9.2).[28]

Triggering Transformation

It's fine to have a vision and ideas for making a better world locally, but how can individuals contribute to the global-scale implementation of the changes that are so urgently needed? Many of the stories in this book are testimony that individual actions in local- and regional-scale stewardship initiatives are critical to success. These initiatives sometimes spread to larger scales. There are also many examples where one or a few individuals triggered efforts that led to large-scale transformational change—such as Wangari Maathai (Chapter 3) and other transformational leaders (Chapter 8). I will add three examples where an extraordinary individual helped overcome major barriers to social and ecological sustainability (poverty, environmental degradation, and politics) to influence national and international outcomes.

Muhammad Yunas was born in a village of what is now Bangladesh.[30] He was a good student and at age 21 received his MA in economics. For 5 years he

Box 9.2 Strategies for Transformation

Transformation of a system creates winners and losers. It has uncertain costs and benefits and requires contentious decisions about allocation between present and future generations. Different groups therefore often disagree about how serious the problems are and whether or how to fix them. Strategy elements that are common to many successful transformations include the following[29]:

Preparing for Transformation

- **Engage a spectrum of allies who are aware of the problems and might be recruited to the cause** (Chapter 8).
- **Identify potential barriers, thresholds, pathways, and triggers for positive change.** Scenario exercises can suggest potential alternative states and ways to get there (Chapter 7).
- **Define a clear goal and steps to achieve it.** Identify barriers to change, potential change agents, and strategies to overcome barriers.

Navigating the Transition

- **Use crises as windows of opportunity to trigger transformation.**
- **Be positive** (Chapter 7). **Seek to attract followers rather than to demonize and overpower the opposition.** Most social movements succeed not by overpowering their enemies but by undermining their support. Desegregation protests in the 1960s began primarily among African Americans but gradually recruited others, including moderate white people who were concerned about segregation.
- **Encourage broad participation and foster communication among organizations and across scales** (Chapter 8). This builds trust and ensures that the process is not co-opted by a small subgroup.
- **Maintain transparent processes and flexible strategies.** Transformation often has unexpected outcomes, so strategies must be nimble and flexible (Chapter 8).

Building Resilience of the New System

- **Build a plan to sustain successes.** Initiate activities that build trust and respect, and identify social values that are held in common (Chapter 8).
- **Create incentives and foster values that support the new context** (Chapter 5).
- **Initiate and mobilize social networks of key individuals for problem-solving** (Chapter 7).
- **Foster interactions and support decision makers at other levels** (Chapter 8).

lectured in economics in Chittagong College near his birthplace, then went to the United States to study economic development. After completing his PhD in 1971, Yunas was active in raising money to support the liberation of his country. After the war, he returned to Bangladesh and joined the planning commission of the new government. However, the job was boring, so he resigned and went back to Chittagong University, where he began research to test various community-based approaches to economic development.

Yunas learned that very small loans made a disproportionate difference in the poorest households. Women who bought bamboo to make furniture had to borrow money from lenders who charged high interest rates. These women were stuck in a poverty trap from which they could not escape. In 1976, he lent US$27 of his own money to each of 42 women, who were able to repay the loans, make a profit, and provide more effectively for their families. Established banks would never make such loans because they were too small to be worth the effort and banks didn't believe that poor women would repay the loans.

Yunas gradually expanded this effort into the Grameen Bank (Village Bank), which by 2017 had 9 million borrowers and a repayment rate of 99.6%. Most (97%) of the loans have gone to women, who suffer disproportionately from poverty and are more likely than men to use their earnings to support their families. Yunas' Grameen microfinance model has inspired similar efforts in about 100 other developing countries and has challenged the pre-vailing economic-development assumption that top-down mega-projects are the best way to bring people out of poverty. This microfinance model earned Yunas and the Grameen Bank the Nobel Prize in Economics. Yunas' accomplishments illustrate the potential of an innovative student and young professional, who began with a small experiment in one rural community and has since transformed the world in significant ways.

Bold stewardship efforts are not always successful. The Great Barrier Reef along the coast of northeast Australia is the largest reef system in the world. It was significantly degraded by the 5- to-10-fold increase in sediment delivery from rivers that resulted from overstocking of ranch lands. Overharvest of fish and climate warming further degraded the reef. An unprecedented warming-induced coral-bleaching event in 1997–1998 was a wake-up call that existing protection efforts were not preventing the deterioration of the reef and that a radical policy change was needed.

In 1998, a few leaders of the Great Barrier Reef Marine Protected Area (GBRMPA) initiated a major transformation of park management to em-phasize the protection of biodiversity and ecosystem services rather than maximizing fisheries yield.[31] The marine park was rezoned to increase the area protected from fishing from 5% to 33%, a step that increased the abundance of some fish species. Unfortunately, the positive effects of innovative management within the GBRMPA were overwhelmed by events that were beyond the park's control. Rapid growth of coastal communities increased fishing pressure. In addition, Australia's decision to develop massive new coal mines inland from the reef required dredging of ports to service large coal-transporting ships. This dredging dumped into the reef an amount of sediment comparable to the natural sediment delivery of all 35 rivers that drain into GBRMPA. Most critically, two additional warming-induced coral-bleaching events (2002 and especially 2016) caused extensive death of corals. The factors that the GBRMPA could control—fishing and water quality—had no effect on reef susceptibility to bleaching. Instead, Australia's decision to massively increase its coal exports directly contributed to two of the greatest threats to the reef: sediment delivery and climate-induced coral bleaching. This illustrates the importance of coherent strategies that link actions at local, national, and global scales.

In 1968, Garrett Hardin wrote a high-profile paper which argued that natural resources held in common by society (the commons), such as rivers, public lands, and the atmosphere, are destined for degradation unless they are either privatized or managed by laws that regulate people's use of resources.[32] This article became a conceptual rallying point for advocates of either privatization of public assets or strict government regulation to control the amount and allocation of rights to graze, fish, or pollute public lands and waters. Since then, Lin Ostrom and others have assembled an impressive body of research showing that, under some circumstances, local communities and societies develop informal rules that effectively regulate use of the commons.[33]

Through study of hundreds of cases where small groups manage their use of common-pool resources, Lin and her colleagues identified the conditions where this informal management approach is most likely to succeed.

- clearly defined boundaries within which the resource can be monitored and managed
- the rights and power of the local group to regulate its own resource use
- equitable distribution of effort and benefits in managing the resource
- low-cost, local arenas in which users can resolve conflicts
- graded sanctions that match the seriousness of offenses

These rules of thumb provide guidance to groups like Maine's harbor gangs (Chapter 8) on ways to improve their management of the commons. This research, for which Lin Ostrom and her colleagues received the 2009 Nobel Prize in Economics, also shows the potential for success of local management of the commons—often without government intervention or privatization of resources. This body of research, like that of Muhammad Yunas, led to a marked change in approach by international development agencies such as the World Bank. They switched from predominant funding of top-down government-managed projects in the 1980s to greater emphasis on bottom-up local management of development.

The success of community management of lands and resources cannot be guaranteed. It depends on local context and leadership and on the rules by which groups collaborate. When the conditions are right, the decisions and behavior of individuals in local settings can foster stewardship outcomes that can be mainstreamed into international policies in ways that are important to the future of the planet. In other words, the stewardship decisions and behavior of individuals can have important global consequences.

What Can We Do?

Most political changes in democracies begin with ideas and actions of individuals—both ordinary and extraordinary. However, people differ in the political roles they choose to play. Options include the following:

- **Fulfill our basic responsibilities as citizens.** Inform ourselves, vote, respect the rights and opinions of others, and encourage everyone to participate fully in democracy.
- **Advocate for stewardship within our social networks.** Inform others of ways to improve the well-being of nature and society. For example, talk to others about climate change—why you care and what you could do together.
- **Reframe debates to address genuine issues rather than generic worldviews.** Issues can be clarified through letters to leaders or newspapers, participation in town-hall meetings, informing others of issues, and leadership of civic groups.
- **Challenge vested interests that undermine stewardship.** Challenge groups that misuse money and power rather than democratic processes to advance their private interests.

The next question is: *If we know how to shape the components of a more sustainable planet, how do we integrate these efforts globally?*

10
Earth Stewardship

Up to this point, I have emphasized actions that individuals can take in support of stewardship. This final chapter expands the focus to consider the interactions of individuals with government, business, and other institutions to build upon and support individual efforts. I emphasize opportunities to achieve sustainable outcomes at regionally and globally significant scales.

Choosing Anthropocene Outcomes

Many geologists contend that Earth has entered a new geologic epoch (the Anthropocene) dominated by human forces that are shaping its future path. The 2017 US *National Climate Assessment* special report points out that "globally averaged air temperature since 1900 is the warmest in the history of modern civilization. Human activities, especially the emission of greenhouse gases, are the dominant cause of the observed warming since the mid-20th century."[1] The atmospheric concentration of carbon dioxide (CO_2), the greenhouse gas that has caused the greatest warming, is now higher than at any time in the history of our species (Chapter 4). These environmental changes interact with direct human impacts on ecosystems in ways that degrade many of nature's services on which society depends (Chapter 3).

If society continues its current pattern of increasing use of fossil fuels, individuals born today will probably experience more warming during their lifetimes (7°F) than has occurred since the beginning of human civilizations 10,000 years ago (about 5°F).[2] The 1%–2% of US and global populations that live in low-lying coastal areas will suffer more frequent coastal flooding due to a combination of rising sea level and more frequent severe storms. Continued increases in droughts, floods, and heatwaves will create additional risks that could trigger mass migrations of *billions* of people—orders of magnitude more than recent migrations to developed nations.[3] These are some of the serious societal consequences of business as usual—that is, choosing *not* to make a concerted effort to reduce human impacts on climate.

Every time I read about a new storm or wildfire that has killed people and caused widespread suffering, I think sadly about the links between climate change and the human consequences of failing to take strong climate action. No single event can be blamed on climate change, but these events will likely continue to increase in frequency and/or intensity until society takes concerted action to reduce human impacts on climate. This is an urgent crisis. The costs in lives and money of delaying strong action for another decade or two are enormous compared to the costs of acting today to reduce human impacts on climate. Other more catastrophic outcomes of climate inaction are possible but uncertain. For example, continued climate warming might tip Earth's climate toward a permanent hothouse state that would render future human decisions about climate inconsequential and would disrupt human civilization as we know it (Chapter 4).[4]

However, these outcomes are not yet inevitable. Society has a choice. Many of the detrimental changes in climate and ecosystems can be slowed or reversed by vigorously pursuing the actions described in this book. After all, fossil-fuel emissions have caused Earth-altering impacts for only about 70 years (Chapter 1). It's not too late to reverse these impacts and therefore reduce the worst consequences of climate warming.[5] However, society needs to initiate concerted actions now. As described in earlier chapters, the following steps would substantially reduce the risks of continued climate warming and environmental degradation and could trigger transformation toward a world in which people and nature can flourish together:

- **Reduce the concentrations of greenhouse gases in the atmosphere.** This requires a reduction in fossil-fuel emissions of CO_2 (Chapter 4). Reducing emissions of other greenhouse gases, such as methane and nitrous oxide that are associated with intensive agriculture, would further reduce human impacts on climate (Chapter 4).
- **Increase the extent of forests and wetlands that naturally remove large quantities of carbon from the atmosphere.** A shift from global deforestation to reforestation and from wetland drainage to restoration would accomplish this goal (Chapter 4).
- **Reduce human population growth.** Improved access to general education for girls in developing nations and to information about family planning for people everywhere—especially young people—would provide immediate societal benefits and reduce growth of the human population (Chapter 3).

- **Sustain the ecosystem services needed by society locally and globally.** Reducing human activities that degrade ecosystems would benefit both people and nature (Chapter 3).

These are not simply climate solutions. Progress toward these goals would provide many direct societal benefits (Figure 4.2). Asking whether or not to take actions that reduce climate change is the wrong question. The key question is: What actions will reduce human impacts on climate most quickly and effectively, while minimizing short-term economic and societal disruption?

Reducing human impacts on the climate system is a daunting prospect because of the global scale and rapid pace at which it must occur. However, we already have a good start. Karen O'Brien argues that the most fundamental transformations needed are changes in the personal attitudes and behavior of individuals. We must each question our own behavior and worldviews and make more intentional individual and collective choices (Chapters 6–9).[6] If widespread, these changes can influence the behavior of others and shift social norms and worldviews toward attitudes that foster sustainability.

The process begins with individual behavior, family habits, and social networks and expands to communities, regions, and nations. In this way, personal transformations provide the foundation for political and social actions that can transform institutions and governments and tip the balance from private vested interests toward broader societal goals (Chapter 9). Technological transformations are also needed. However, most technologies needed to rein in climate change are already advanced enough to implement now, although additional research would increase their cost-effectiveness and efficiency (Chapter 4).

Fostering Pursuit of Happiness

In 1776, the US Declaration of Independence radically identified "pursuit of happiness" as an **inalienable right** of all people (that is, a right that cannot be taken away), alongside life and liberty.[7] Social scientists can now measure happiness and life satisfaction and therefore have a reasonably good understanding of what makes people happy. Among poor people, income is a good predictor of happiness because it increases people's capacity to meet their basic needs. However, among the 85% of Americans who do not live in poverty, there is no relationship between income and people's self-reported happiness (Chapter 6).

In the last two and a half centuries, an economic system has developed that has emphasized the generation of individual wealth rather than pursuit of happiness (see next section). There is now a dual imperative to restore happiness as a core national goal: (1) the pursuit of material wealth, which has accelerated since the mid-20th century, has failed to increase the happiness of most people living in the United States and other developed nations (Chapter 6) and (2) pursuit of wealth has degraded Earth's capacity to deliver nature's services on which society depends (Chapter 3).

Bhutan is a small Buddhist country between India and Tibet. It has specified happiness rather than wealth as a core national goal.[8] Bhutan incorporated **gross national happiness (GNH)** into its constitution in 2008 as it moved from monarchy toward democracy. Some people viewed GNH as a smokescreen to draw attention away from Bhutan's civil-rights violations and genocide of Nepali refugees. Others see GNH as an effort by a fledgling democratic government to distance itself from past government policies.

The GNH index, as developed in Bhutan, has four pillars: fair socioeconomic development (better education and health), promotion of a vibrant culture, environmental protection, and good **governance** (the processes of governing). These four elements are similar to the foundations of well-being that underlie Maslow's list of fundamental human needs (Chapter 5).

GNH has captured the imagination of scholars and communities in many places around the world, just as the goals of American democracy—life, liberty, and pursuit of happiness—inspired other nations to move toward democracy 250 years ago.

In 2011, the United Nations General Assembly urged all nations to measure happiness and promote it as a fundamental human goal.[9] New indices have been developed that provide a more comprehensive measure of well-being than does gross domestic product (GDP). The genuine progress indicator, for example, developed by ecological economists, includes environmental, social, and economic dimensions of well-being.[10] A renewal of society's commitment to *pursuit of happiness* by people living in democracies would go a long way toward fostering a world in which people and nature could flourish together. Let me explain.

If pursuit of happiness is an inalienable human right, society must do the following:

- **Ensure that all people can meet their basic needs for food, shelter, health, and safety** (Chapter 5). International humanitarian aid programs channel aid to developing nations, and national and local governments seek to meet these needs for their citizens. When government safety nets are insufficient, family, friends, and community organizations are the only options to fill the gap. The logical individual actions are to (1) pressure government to support welfare programs for vulnerable people and

(2) participate in community organizations that supplement this safety net when government support is insufficient.

- **Provide educational opportunities that sustain culture and empower people to shape their own lives** (Chapter 5). Informal education through social interactions with family, friends, and community groups sustains cultural values. Formal education provided by government conveys additional knowledge and skills that are less deeply embedded in culture. Businesses and agencies often supplement public education by providing specialty training for their employees. Support for both government-funded public education and community learning and enrichment opportunities empowers people to achieve their goals.

- **Constrain environmental impacts within limits that conserve ecosystem services** (Chapter 3). Governments regulate pollution and other activities that directly threaten the health and well-being of people and other species. Sustaining this environmental protection requires public advocacy. However, there is no strong mechanism to ensure that regulations are sufficient to meet the needs of future generations. This responsibility often falls to environmental groups supported by public advocacy or to individuals and government agencies that manage their own lands.

- **Reduce the consumption of resources that degrade ecosystems but fail to increase people's happiness and life satisfaction** (Chapter 6). Ultimately, changes in consumption behavior are the responsibility of individual consumers. In addition, businesses sometimes promote sustainability through environmentally friendly practices that they advertise to consumers. Government can also influence consumer choices through subsidies and incentives.

- **Encourage government to focus on the needs of all of society rather than privileging vested interests** (Chapter 9). Curbing excessive influence of vested interests on governance is most effectively addressed through political processes. These include voting, communicating with political representatives, engaging with organizations that promote democratic processes, shaming entities that violate democratic principles, and participating in nonviolent political protests.

Harnessing the Economy for Stewardship

The only substantive argument that has been raised against pursuit of stewardship and sustainability is that it might weaken national or global economies. However, Naomi Klein, in her book *This Changes Everything: Capitalism vs the*

Climate, argues that the capitalist economy is the root cause of both climate change and many of the social and economic ills currently faced by global society.[11] She correctly points out that some of the greatest barriers to climate action are vested interests in business and politics that both benefit from and promote today's energy-intensive economy. She therefore advocates political action to confront these interests as *the* critical step needed to reduce rates of climate change. I agree that political action is essential in a democracy—especially when powerful vested interests operate outside of democratic processes (Chapter 9). I would argue, however, that the market system doesn't need to be scrapped. Indeed, market forces can exert a potentially *positive* role by promoting rather than undermining the needed transformation toward sustainability. Let me explain.

The classical model of market economies is based on the behavior of a fictional species—*Homo economicus* (rational economic individuals).[12] In this model, consumers make rational choices that reflect their social and material preferences. They strive to meet their greatest desires and needs by buying the material necessities, conveniences, and luxuries that they most want, as well as by pursuing actions that reflect their values toward family, friends, society, and nature. Economists assume that people's observed choices are those that give them the greatest satisfaction, within the limitations of what they know or believe. Businesses maximize profit by producing the goods and services that consumers are most willing to buy, so their decisions also reflect people's individual choices.

Economists recognize that many goals contribute to people's satisfaction as they decide how to spend their time and money and that different people rank these goals differently. These goals range from total materialism to the satisfaction of acting in ways that benefit society, nature, and future generations. As extreme caricatures, let's imagine two types of individuals: *Homo materialis*, who is motivated primarily to buy and consume material goods, and *Homo environmentalis*, whose actions and choices reflect greater concern for the welfare of society and the environment—today and in the future. Real people in society span the spectrum between these two extreme caricatures. Most of us buy and consume more material goods than we really need (Chapter 6), but we also make choices that reflect our care and empathy for other people and for nature (Chapter 5).

Steve Polasky, an ecological economist, pointed out to me that the current economic system was spectacularly successful at meeting the major challenge that society faced at the beginning of the Industrial Revolution (or even from 1900 to 1950), when material scarcity deprived many people of their basic needs for a satisfying life.[13] This economic system focused primarily on providing material goods and motivating people through advertising to choose which goods to buy. This emphasis on supplying and consuming material goods (the *Homo materialis* caricature) creates profits for businesses that sell the goods preferred by society. Businesses encourage government to formulate policies that promote individual consumption and corporate profit—which are reflected at the national scale as pressures for policies to increase GDP.

Material scarcity is still a defining characteristic for many people—especially in developing nations—but this is *not* the main problem for the developed world. As a global society, the main problem we face today is the degradation of Earth's environment at the planetary scale.[14] This degradation is the unintended consequence of economic growth that is substantially based on resource extraction. The drivers of planetary degradation include resource consumption, population increase, deforestation, climate warming, biodiversity loss, and many other earth-changing processes (Figure 10.1).

Figure 10.1 Pogo reflecting on the causes of environmental problems, as pictured on the first Earth Day poster.

The crucial question is how to shift from a system aimed at solving the problem of producing sufficient quantities of goods to one that motivates stewardship of social-ecological systems in ways that allow people to lead satisfying lives—both today and in the future. Imagine a world in which people's consumption choices are substantially influenced by their societal and environmental values (the *Homo environmentalis* caricature; Chapter 6). Imagine that attitudes based on these values spread through society by intentionally shaping social norms, encouraged by religious and other community groups. Imagine also that government policies—advocated by citizens—foster environmentally favorable choices (Chapter 9). The market would then motivate businesses to make products that match these environmental preferences (Chapter 6).

When Swiss citizens were allowed only a small weekly bundle of nonrecyclable garbage, they preferred products with less packaging; and businesses reduced their packaging to match these consumer preferences (Chapter 6). How much would society suffer if products were not encased in single-use plastics that are immediately thrown away? Likewise, high gasoline prices during the oil crisis of the 1970s motivated people to buy smaller, more fuel-efficient cars. This led car manufacturers to make these types of cars, and advertisers touted the virtues of fuel efficiency rather than horsepower. Businesses and advertisers are certainly motivated to sell as much stuff as possible. However, consumers need not be mindless pawns. With intention and coordination, people can shape markets significantly through both consumer choice and support for appropriate government policies.

Reshaping the economy to reduce its environmental impacts need not destroy it. It would likely shift jobs from production of fossil fuels to production of renewable energy, from resource extraction to resource recycling, from timber management to forest management for a broader mix of ecosystem services, including timber, water, recreation, and climate protection. In the short term, there would be both winners and losers. Reducing the hardships and encouraging new opportunities that come with economic transitions is a genuinely important challenge for government, business, and civil society; but it is not a reason to continue degrading planet Earth to the detriment of most of humanity.

In the end, the market economy is largely a reflection of consumer choices. These choices are sometimes short-sighted, if people are unaware of the impacts of purchasing a particular product or are driven by status to purchase a bigger house or fancier car than they need (Chapter 6). These choices that do not contribute to personal satisfaction (that is, **market imperfections**) can be shaped by thoughtful and intentional behavior of consumers and policies of government, as described throughout this book. In other words, individuals ultimately determine whether market forces promote or undermine sustainability.

Economists and politicians argue passionately about how the economic system should be shaped to improve society's well-being and happiness. Some economists and politicians have great faith in markets to deliver good outcomes for society. Others see an important role for government in minimizing situations where unregulated markets reduce societal well-being (**market failures**). Regardless of people's economic and political perspectives, there are several economic and societal actions that could foster prosperity with less dependence on the overconsumption and economic growth that degrade our planet:

- **Incorporate costs of environmental damage into costs borne by business—the polluter-pays principle.** This well-accepted principle of resource economics uses government regulations to define and enforce the pollution limits that society considers appropriate (Chapter 8). Application of this approach in the United States in the 1970s rapidly reduced water and air pollution and acid rain. If similar regulations set a price on carbon emissions, industry would have strong incentives to develop and implement cost-effective low-carbon energy technologies (Chapter 8).
- **Ensure that wages meet the basic material needs of poor families** (Chapter 5). The US federal minimum wage has declined in real terms for half a century (since 1968).[15] In 2018, a person working fulltime at the US minimum wage ($7.25/hour) received an income ($15,000) that was less than the poverty threshold for a family of two ($16,000). People without full-time work earn less than this, and people with larger families or greater debt have additional costs. Thirty states and many cities, such as San Francisco, New York, and Chicago, have set higher minimum wages without experiencing economic hardship.[16]
- **Invest in people and jobs** (Chapter 5). Businesses often provide training that increases the productivity of their skilled workers. Government job-training programs are important for less skilled jobs, especially in places where industries have closed. This job training could target professions (such as green-energy jobs or people who maintain electrical grids), where technological change or impending retirements greatly increase the need for more workers.
- **Reduce debt-driven consumption.** Rising household debt over the last three decades has increasingly fueled consumption and propped up

the US economy, at a time when the size of the middle class has shrunk (Chapter 5).[17] Although debt-supported consumption boosts corporate profits, it has destabilized the American economy and increased the financial risks to individual citizens. A combination of financial regulations and corporate and civic responsibility could reduce household debt and associated consumption and provide greater financial security and self-satisfaction to citizens.[18]

Engaging in Political Change

Many stewardship opportunities benefit from a mix of individual, business, and government actions. Reducing rates of climate change can happen through reduced energy use by individuals and climate-friendly policies of governments at local to global scales. Innovation and investment by industry can also reduce climate impacts by providing and using less carbon-intensive forms of energy. I firmly believe that choices made by individuals can influence not only their own individual and collective behavior but also the actions of business (through consumer choices) and government (through active engagement in political processes).

Government, in turn, influences individual stewardship actions through laws and regulations that make it easier for citizens to behave sustainably, as described in earlier chapters. At the local level, for example, conservation easements make it easier for individuals to sustain working lands rather than selling them for commercial development (Chapter 6). Bicycle- and pedestrian-friendly paths and corridors make it easier for individuals to reduce their use of fossil fuels. At larger scales, government can reduce pollution and improve public health by raising fuel-efficiency standards, improving public transportation, and putting a price on carbon emissions.

Government is unlikely to foster social and environmental sustainability unless voters expect and demand these actions (Chapter 9). Voting is therefore critical in fostering stewardship by government, especially when voter turnout is low, as is often the case in the United States. Some elections in my town are decided by fewer votes than the number of people I might encourage to go to the polls on election day. My preferred candidate for the local electric utility board won by five votes (June 2018), increasing board support for renewable energy. My preferred candidate for the Alaska House of Representatives lost to her opponent by one vote (November 2018), making me wish I'd put greater effort into her campaign.

Even in US national elections, turnout of eligible voters has averaged only about 60% in presidential elections and 40% in midterm elections over the past century.[19] Another 25% of potential voters haven't even registered to vote. Since 1990, presidential elections have been decided on average by less than 5% of the popular vote. That's a small enough proportion of the electorate that concerted public support and voting for the well-being of society and health of ecosystems could profoundly enhance government advocacy for stewardship in a single election cycle.

Surveys of people who don't vote show that people say they are too busy or aren't interested.[20] There are many ways to make voting easier. Thirteen states automatically register citizens when they obtain or renew their driver's license or state identification card, and ballots are mailed to registered voters after their 18th birthday. Public disinterest can also be overcome through voter information campaigns and by conversations within social networks about the importance of voting—talk to your neighbors and friends! The flip side of low voter turnout is that those people who do vote have a disproportionately large impact on the outcome of elections. That's all the more reason that each person who cares about the future should vote for it.

However, according to Pew Research Center findings in 2015, public trust in the US government is at an all-time low.[21] Citizens are also paralyzed by their distrust of organized labor, big business, public schools, television, and even their leaders. Robert Putnam, in his book *Bowling Alone: America's Declining Social Capital*, argues that, since 1950, the United States has been moving in the wrong direction. Americans have been disengaging and losing their capacity to come together to solve problems.[22] Membership in American civic groups has declined as people turn inward and focus on video games and virtual unrealities rather than engaging with others to address real problems. Yet there have certainly been times, as noted earlier, when Americans did come together to address social and environmental issues.

In my view, the degeneration of public debate into clashes between generic political worldviews has become a smokescreen that distracts people from the real issues facing society (Chapter 9). In the resulting political gridlock and public disengagement, politicians maneuver to consolidate their power and "win" rather than coming together to solve problems. Under these circumstances, it is easy for vested interests to hide behind generic worldviews rather than addressing factual issues. It is then difficult for government to provide leadership on behalf of humanity and the environment. In the past, social and environmental crises sometimes triggered collective public actions that pressured government to address critical issues. These concerns included women's right to vote in the 1910s, racial equality in the 1960s, and environmental protection in the 1970s. These initiatives arose mainly from society rather than from entrenched political leadership. I believe that government action on behalf of the rights of the planet and its citizens is most likely when public dialogue, voting, and actions demonstrate public support for the well-being of society and the rest of nature. Only under these circumstances are governments likely to provide leadership.

Globalizing an Ethic of Care

The fundamental human instinct to compete empowers people to meet many of their basic needs. This competitive instinct leads *Homo materialis* to pursue wealth and power, in part by acquiring as many necessities, conveniences, and luxuries as possible. These competitive instincts are fueled by advertising and by institutions such as the stock market that can create wealth or loss and by credit cards that put debt within anyone's reach.

Yet, people also have an instinct to care. I argue that the instinct to care for others and for the environment is important in meeting Maslow's entire spectrum of human needs over the long run (Chapter 5). When people share and help others, poor or vulnerable people are better able to meet their basic needs, and all people are more likely to develop the trust and friendship that underlie good social relations. Trust and friendship, in turn, nourish people's motivation and capacity to meet their personal, spiritual, and societal goals.

The globalization of human interactions extends the capacity of people to care for others and for the environment. Governmental, religious, and community groups provide safety nets that enable people to meet and care for one another. The major 20th-century social movements in the United States (for example, the rights of people regardless of race, gender, and sexual preference and the rights of the environment) are based on empathy and humanity's capacity to care. Greater protection of these rights often happened quickly once there was open public support. International support for the United Nations' Millennium Development Goals during the early 21st century shows that institutions based on caring can make a big difference at the global scale in enabling people to pursue health and happiness (Chapter 5).

In some Western countries, such as the United States and the United Kingdom, government social programs that care for vulnerable people and the environment have been gradually eroded as people and governments emphasize economic competition as a way to solve social problems. This shift from policies based on caring to policies based on competition gives rise to at least two pathways by which caring instincts can foster the pursuit of happiness. Political action by society can pressure governments to fund and restore their responsibility to protect their citizens' inalienable right to the pursuit of happiness. At the same time, participation in nonprofit groups focused on social and environmental justice allows people to extend their care to people and environments they know well, as well as to others whom they may never meet.

Within Our Reach?

The enormity of global problems is genuine cause for concern about the future of our planet. Deep despair leads people to disengage from problems rather than coming together to solve them (Chapter 7). However, there is substantial evidence that society *has the capacity* to take stewardship actions with regionally and globally significant impacts. Following are some examples:

- Research has shown that changes in land use and climate are the major causes of global environmental degradation and has identified actions that would reverse these impacts. This provides clear rationale and guidance for developing globally effective stewardship strategies.
- A large proportion of Americans (40%–80%) self-identify as environmentalists. An estimated 2 million groups worldwide actively address social and environmental equity and sustainability. This is perhaps the largest social movement in the history of humanity (Chapter 8). This shows widespread public concern for the future of vulnerable people and the environment.
- A global network of 1,500 cities, towns, and regions that serve 25% of the global urban population has committed to a sustainable future by developing green, resource-efficient economies (Chapter 1). This shows widespread public commitment to stewardship action.
- Collaborative resource management has emerged in many regions from either shared or contested visions for a region's future (Chapter 8). Deep political divides do not necessarily preclude stewardship.
- Government regulations in many developed nations have reduced acid rain, air and water pollution, and ozone depletion (Chapter 3). This shows the potential for significant stewardship outcomes.
- Transnational corporations representing 10% of global GDP have designed strategies to sustainably harvest raw materials and still be profitable (Chapter 1). This shows the potential of business to contribute to global stewardship.

- The United Nations has identified 17 sustainable development goals to transform toward a more sustainable future. These goals guide the funding decisions of many national and international development agencies. Within 25 years (by 2015), these efforts had already substantially reduced global poverty and disease and increased educational opportunities for poor people (Chapter 5). This shows the potential for global collaboration in support of stewardship.

These accomplishments are insufficient, but they provide evidence that stewardship outcomes are within our reach at local to global scales.

Unshrouding the Clouded Crystal Ball

Although society faces daunting local and global problems, we collectively have much of the knowledge, tools, and wisdom needed to address them. Most of the individual steps are not difficult, although the pathway is complex and challenging.

Let me share an allegory about walking toward a mountain. As the ancient Chinese philosopher Laozi (Lao Tsu) pointed out, "a journey of a thousand miles begins with a single step"—an intentional one.[23] When I start this journey, the first step is the most difficult. I may not see the need to take this step or see the relationship of the first step to my distant mountain goal. Besides, the first step seems so inconsequential. However, the first step moves my body forward. My other foot must then reach out to keep my body balanced, so the second step is easier. And then the third. As I walk, some paths lead more directly to my mountain goal or are easier because they fit my skills and passions. Along the path, I meet other people, and we become friends of circumstance, regardless of where we each began. We share stories, learn from one another, and adjust our plans and routes accordingly. A few may have begun their journey with a map and a clear destination in mind. Others began with an afternoon stroll and became caught up in their own inspiration or the visions of others.

As I near the mountain, I see that its green cloak is thinner and grayer than I remember, and little creatures are nibbling at its heels. It has aged before its time. The mountain tells its story as we compare notes, and we imagine how we might become closer friends—each helping the other go where we can flourish together.

This allegory suggests many different levels of commitment to stewardship and multiple pathways and goals. Although the crystal ball is clouded, there are many tools and strategies that can increase the likelihood of success.

In May 2016, King Carl Gustaf of Sweden assembled 24 researchers in a royal colloquium to rethink options for a sustainable future for planet Earth.[24] Each participant came with a distinct interest and experience—for example, energy futures, factors tipping the balance between peace and war, human indifference to climate change, triggers for transformation to sustainability. This is just one of many conferences where citizens, scientists, policymakers, or managers have worked together to identify practical steps toward stewardship. King Carl Gustaf's decision to organize a dozen royal colloquia over 25 years did a great deal to advance the science and practice of sustainability and create a Swedish national commitment to this goal. Although most of us are not kings, each of us has a unique set of passions, skills, and networks that can help shape our planet's future in ways that no one else can do.

I don't honestly expect society to achieve perfect stewardship, nor do I believe that such an outcome exists, given the constantly shifting currents and eddies of change. However, I'm certain that society can shape a much more promising path toward its own well-being and the health of ecosystems if people act with intention and inspire others to share these pathways. Although many of humanity's big opportunities require concerted collaboration of individuals, business, and government, none of this will happen without active engagement and guidance by individual members of society. If individual people don't speak and act on behalf of nature and society, neither government nor business will be motivated to take the lead. On the other hand, if people pursue their own rights to long-term happiness and the rights of others and of nature, government and business are more likely to be willing partners, and widespread transformation can occur.

My grandchildren Adele and Emile, who inspired me to write this book (Prologue), usually walk, bicycle or take public transport to school and voice their concern about the future by participating in climate marches. Although they are only 13 and 10, respectively, they want to do their part to make the world a better place. If people decide, individually and collectively, that our planet's future habitability is important, we can ensure a better life for our grandchildren and their grandchildren's grandchildren. We can't sit back and wait for government or someone else to take the first step. It's up to each of us—now! Besides, it's the right thing to do.

Acknowledgments

I thank the many students, colleagues, and other people who contributed to my understanding of stewardship. Among these mentors, I particularly thank Elaine Abraham, Neil Adger, Catherine Attla, Todd Brinkman, Katrina Brown, Steve Carpenter, Mimi and Stuart Chapin, Jr., Patricia Cochran, Bill Denison, Val Eviner, Carl Folke, Buzz Holling, Henry Huntington, Orville Huntington, Jill Johnstone, Charlie Jones, Corrie Knapp, Gary Kofinas, Amanda Lynch, Pamela Matson, Hal Mooney, Kathleen Dean Moore, Robin Reid, AlexAnna Salmon, Marten Scheffer, Gus Shaver, Monica Turner, Peter Vitousek, and Brian Walker. I also thank my grandchildren, Adele and Emile Chapin, who inspired me to write this book.

Many groups and institutions provided the intellectual ferment and camaraderie that pushed me out of the ivory tower and beyond my comfort zone to seek stewardship solutions. These groups include the Resilience and Adaptation Program, the Community Partnership for Self-Reliance, and the Bonanza Creek LTER program at the University of Alaska Fairbanks; the Resilience Alliance, the Stockholm Resilience Center, and the Beijer Institute of Ecological Economics at Stockholm University; the Ecological Society of America; and the Fairbanks Climate Action Coalition.

People who generously shared stories and experiences and reviewed my renditions of these stories include Stephan Barthel; Charley Basham; Elena Bennett; Robin Bronen; Caroline Brown; John Bryant; Shauna BurnSilver; Steve Carpenter; Adele, Keith, and Mimi Chapin; Patricia Cochran; Johan Colding; Laʼona DeWilde; Jessica Girard; Brad Griffith; Katharine Hayhoe; Orville Huntington; Aliosha Imnosentsva; Torre Jorgensen; Gary Kofinas; Jim Magdanz; Alan Mark; Mila Kallen Marshall; Pamela Matson; Kevin McAleese; Larry Merculieff; Moreen Miller; Hal Mooney; Bill Nell; Stephanie Quinn-Davidson; AlexAnna and Christina Salmon; Martha Shulski; Ben and Carrie Stevens; Bill Streever; Peter Vitousek; Cathy Walling; and Tom Webb.

Others who provided valuable feedback on portions of the manuscript include Neil Adger, Dorothy Boorse, Katrina Brown, Karen Colligan-Taylor, Paul Ehrlich, Erle Ellis, Eugenie Euskirchen, Hannah Gosnell, Josh Greenburg, Hal Harvey, Alison and Jim Henderson, Greg Hitzhusen, Rich Hum, Jill Johnstone, Joe Little, Dave McGuire, Alison Meadow, Ted Moore, Roz Naylor, Michael Nelson, Dan O'Neill, Sara Peacock, Steve Polasky,

Carmen Revenga, Brian Rogers, Clara Schuur, Heather Tallis, Kristin Timm, Jeroen van den Bergh, John Walsh, and Marge Zielinski.

I particularly thank Aleya and Todd Brinkman, Joan Canfield, Steve Carpenter, Mimi Chapin, Val Eviner, Carl Folke, Corrie Knapp, Carolyn Kremers, Jeremy Lewis, Dave Long, Raphaël Mathevet, Hal Mooney, Kathleen Dean Moore, Joanna Nelson, Marten Scheffer, Cam Webb, Tom Webb, Stan Zielinski, and five anonymous reviewers who provided insightful suggestions and feedback on the entire draft manuscript. Finally, I thank my son Mark, who helped me with photos and drafted the cover design, and my wife Mimi, who has been my toughest critic and staunchest supporter throughout the writing process, as well as crafting the line drawings.

Published Guides for Climate Action

Energy Conservation

Calculating your carbon footprint[1]
Conserving energy[2]
Increasing energy efficiency[3]
Renewable energy and independent living[4]

Honorable Consumption

Reducing unnecessary consumption[5]
Simple ways to reduce food waste[6]
Assessing product quality[7]

Green Labeling and Socially Responsible Investment

Green labeling of products[8]
Life-cycle analysis[9]
Environmentally responsible investment[10]

Volunteering for a Healthy Environment

Volunteering for sustainability[11]

Fostering Community Change

Principles for engagement[12]
Communicating respectfully about realities of climate change[13]
Collaborating to develop solutions[14]
Just transitions[15]
Triggering transformation[16]

Comprehensive Strategies for Reversing Climate Change

The four most effective actions to reduce climate change[17]
A comprehensive plan to reverse global warming[18]
25 easy steps to protect the climate[19]
Living better for you and the climate[20]
50 ways that you and your church can help save the Earth[21]
Empowering kids for climate action (and getting a good night's sleep)[22]
Inspiring solutions: Seeds of good Anthropocene outcomes[23]

References to Climate-Action Guides

1. Environmental Protection Agency. July 14, 2016. Carbon footprint calculator. https://www3.epa.gov/carbon-footprint-calculator/; Global Footprint Network. 2020. Ecological footprint. https://www.footprintnetwork.org/our-work/ecological-footprint/.

2. Conserve Energy Future. 151 ways to save energy. https://www.conserve-energy-future.com/151-ways-to-save-energy.php; WikiHow Jan. 22, 2020. How to conserve energy. https://www.wikihow.com/Conserve-Energy.

3. Energy Star. 2020. Energy star most efficient 2020. https://www.energystar.gov/products/most_efficient/products.

4. Kemp, W.H. 2009. *The Renewable Energy Handbook: The Updated Comprehensive Guide to Renewable Energy and Independent Living.* Tamworth, ON, Canada: Aztext Press.

5. Story of Stuff. 2018. https://storyofstuff.org/movies/story-of-stuff/.

6. Wikiel, Y. April 15, 2019. How to reduce food waste in your home. Real Simple. https://www.realsimple.com/food-recipes/shopping-storing/food-waste-in-america.

7. Consumer Reports. 2020. https://www.consumerreports.org/cro/index.htm.

8. Wikipedia. Feb. 13, 2020. Ecolabel. https://en.wikipedia.org/wiki/Ecolabel.

9. Environmental Literacy Council. Life Cycle Analysis. https://enviroliteracy.org/environment-society/life-cycle-analysis/.

10. Howard, E. June 23, 2015. A beginner's guide to fossil fuel divestment. *The Guardian.* https://www.theguardian.com/environment/2015/jun/23/a-beginners-guide-to-fossil-fuel-divestment.

11. Greening of Detroit. Growing tomorrow's Detroit. https://www.greeningofdetroit.com; Mansfield, C. May 14, 2009. 21 places to look for green volunteering opportunities. GreenBiz. https://www.greenbiz.com/blog/2009/05/14/21-places-look-green-volunteering-opportunities.

12. Jemez Principles for Environmental Organizing. Dec. 1996. Southwest Network for Environmental and Economic Justice. http://www.ejnet.org/ej/jemez.pdf; Environmental Justice/Environmental Racism. Key documents. Energy Justice Network. http://www.ejnet.org/ej/.

13. Gillis, J. 2019. Climate change is complex. We've got answers to your questions. *New York Times.* https://www.nytimes.com/interactive/2017/climate/what-is-climate-change.html; Corner, A., C. Shaw, and J. Clarke. Jan. 2018. *Principles for Effective Communication and Public Engagement on Climate Change: A Handbook for IPCC Authors.* Oxford, UK: Climate Outreach. https://wg1.ipcc.ch/AR6/documents/Climate-Outreach-IPCC-communications-handbook.pdf; Quaker Earthcare Witness. Talking about climate change: Two new guides. https://www.quakerearthcare.org/article/talking-about-climate-change-two-new-guides.

14. Wondolleck, J.M. and S.L. Yaffee. 2000. *Making Collaboration Work: Lessons from Innovation in Natural Resource Management.* Washington, DC: Island Press. Ansell, C. and A. Gash. 2008. Collaborative governance in theory and practice. *Journal of Public Administration Research and Theory* 18(4):543–571.

15. Swilling, M. and E. Annecke. 2012. *Just Transitions: Explorations of Sustainability in an Unfair World.* Tokyo: United Nations University Press; Wikipedia. Feb. 10, 2020. Just Transition. https://en.wikipedia.org/wiki/Just_Transition.

16. Satell, G. and S. Popovic. Jan. 27, 2017. How protests become successful social movements. *Harvard Business Review.* https://hbr.org/2017/01/how-protests-become-successful-social-movements.

17. David Suzuki Foundation. 2020. Top 10 things you can do about climate change. https://davidsuzuki.org/what-you-can-do/top-10-ways-can-stop-climate-change/;   McMahon, J. July 13, 2017. The four most effective things you can do about climate change, according to science. *Forbes.* https://www.forbes.com/sites/jeffmcmahon/2017/07/13/the-four-most-effective-things-you-can-do-about-climate-change-according-to-science/#7ae7aa8c635c; Phys Org July 11, 2017. The most effective individual steps to tackle climate change aren't being discussed. https://phys.org/news/2017-07-effective-individual-tackle-climate-discussed.html.

18. Hawken, P. (editor). 2017. *Drawdown: The Most Comprehensive Plan Ever Proposed to Reverse Global Warming.* New York, NY: Penguin Books.

19. Climate Change. Sept. 29, 2016. What you can do about climate change. US Environmental Protection Agency. https://19january2017snapshot.epa.gov/climatechange/what-you-can-do-about-climate-change_.html.

20. Albeck-Ripka, L. How to reduce your carbon footprint. *New York Times* https://www.nytimes.com/guides/year-of-living-better/how-to-reduce-your-carbon-footprint?campaignId=74XU8.

21. Barnes, R.J. 2016. *50 Ways to Help Save the Earth: How You and Your Church Can Make a Difference.* Louisville, KY: Westminster John Knox Press.

22. DeMocker, M. 2018. *The Parents' Guide to Climate Revolution: 100 Ways to Build a Fossil-Free Future, Raise Empowered Kids, and Still Get a Good Night's Sleep.* Novato, CA: New World Library.

23. Seeds of good anthropocenes. 2019. https://goodanthropocenes.net/other-collections-of-great-ideas/.

Notes and References

I emphasize a mix of academic literature and Internet references that are interesting, accessible, and factually based. I also insert comments about points that have been debated in the academic literature as a guide to interesting issues that are beyond the scope of this book. All Internet links were last accessed in February 2020.

The COVID-19 pandemic, which coincided with the final stages of proofreading this book, radically repurposed the Internet and public access to electronic information sources. This makes the longevity uncertain for the URLs associated with references cited in this book. In most cases, the cited references can also be obtained in print or electronic media from the publisher or organization listed with each reference.

Prologue: My Journey to This Book

1. Caldbick, J. 2012. *Celilo Falls Disappears in Hours After the Dalles Dam Floodgates Are Closed*. Feb. 10, http://www.historylink.org/File/10010.
2. UNICEF. 2018. *UN Convention on the Rights of the Child*. https://www.unicef.org/rightsite/files/uncrcchilldfriendlylanguage.pdf.

Chapter 1

1. Parsons, D. 2019. *Cyanobacteria & Oxygen. Wonder Science*. March 12, https://wonderscience.com/latest/the-great-oxygenation-event/.
2. Ibid.
3. Lenton, T.M. et al. 2016. Earliest land plants created modern levels of atmospheric oxygen. *Proceedings of the National Academy of Sciences* 113(35):9704–9709. https://doi.org/10.1073/pnas.1604787113.
4. Steffen, W. et al. 2011. The Anthropocene: From global change to planetary stewardship. *Ambio* 40(7):739–761. https://www.ncbi.nlm.nih.gov/pmc/articles/PMC3357752/.
5. Folke, C. et al. 2011. Reconnecting to the biosphere. *Ambio* 40(7):719–738. https://www.ncbi.nlm.nih.gov/pmc/articles/PMC3357749/.
6. Huntington, H.P. et al. 2006. The significance of context in community-based research: Understanding discussions about wildfire in Huslia, Alaska. *Ecology and Society* 11(1):40. http://www.ecologyandsociety.org/vol11/iss1/art40/.
7. Watt-Cloutier, S. 2015. *The Right to Be Cold: One Woman's Story of Protecting Her Culture, the Arctic and the Whole Planet*. Toronto, Canada: Penguin Canada Books.
8. Leopold, A. 1949. *A Sand County Almanac*. Oxford, UK: Oxford University Press.
9. Sand County Foundation. https://sandcountyfoundation.org.

10. Mitsch, W.J. et al. 2001. Reducing nitrogen loading to the Gulf of Mexico from the Mississippi River basin: Strategies to counter a persistent ecological problem: Ecotechnology—the use of natural ecosystems to solve environmental problems—should be a part of efforts to shrink the zone of hypoxia in the Gulf of Mexico. *BioScience* 51(5):373–388. https://academic.oup.com/bioscience/article/51/5/373/243987.

11. Sand County Foundation.

12. Chapin, F.S., III, G.P. Kofinas, and C. Folke (editors). 2009. *Principles of Ecosystem Stewardship: Resilience-Based Natural Resource Management in a Changing World.* New York, NY: Springer.

13. Barthel, S., C. Folke, and J. Colding. 2010. Social-ecological memory in urban gardens—Retaining the capacity for management of ecosystem services. *Global Environmental Change* 20(2):255–265. https://doi.org/10.1016/j.gloenvcha.2010.01.001.

14. Barthel, S. et al. 2005. History and local management of a biodiversity-rich, urban cultural landscape. *Ecology and Society* 10(2):10. http://www.ecologyandsociety.org/vol10/iss2/art10/.

15. Ibid.

16. Ibid.

17. Hopkins, R. and L. Astruc. 2017. *The Transition Starts Here, Now and Together.* Paris, France: Actes Sud. https://transitionnetwork.org.

18. Ibid.

19. ICLEI—Local Governments for Sustainability. http://iclei.org.

20. World Business Council for Sustainable Development. https://www.wbcsd.org.

21. Pope Francis. 2015. *Laudato Si',* The Vatican. http://w2.vatican.va/content/francesco/en/encyclicals/documents/papa-francesco_20150524_enciclica-laudato-si.html.

22. United Nations. 2019. *Millennium Development Goals.* https://www.un.org/millenniumgoals/2015_MDG_Report/pdf/MDG%202015%20rev%20(July%201).pdf .

23. Folke et al. 2011. Reconnecting to the biosphere.

24. Editors. Jan. 29, 2020. Phaethon. *Encyclopaedia Britannica.* https://www.britannica.com/topic/Phaethon-Greek-mythology.

25. Steffen et al. 2011. The Anthropocene; Ellis, E.C. 2018. *Anthropocene: A Very Short Introduction.* New York, NY: Oxford University Press.

26. Chapin et al. 2009. *Principles of Ecosystem Stewardship*; Chapin, F.S., III, P.A. Matson, and P.M. Vitousek. 2011. *Principles of Terrestrial Ecosystem Ecology.* 2nd edition. New York, NY: Springer.

27. DeFries, R. 2014. *The Big Ratchet: How Humanity Thrives in the Face of Natural Crisis.* New York, NY: Basic Books.

28. Folke et al. 2011. Reconnecting to the biosphere.

29. Roser, M., H. Ritchie, and E. Ortiz-Ospina. May, 2019. *World Population Growth.* https://ourworldindata.org/world-population-growth.

30. Global Footprint Network. 2019. *Ecological Footprint.* https://www.footprintnetwork.org/our-work/ecological-footprint/.

31. Golux, S. 2010. Envisioning the future. *In a Time of Change.* https://itoc.alaska.edu/previous-projects/.

32. Thunberg, G. 2018. The disarming case to act right now on climate. TED Talk. https://www.ted.com/talks/greta_thunberg_the_disarming_case_to_act_right_now_on_climate#t-656015. https://www.youtube.com/watch?v=VFkQSGyeCWg.

33. Xu, C. et al. In press. Future of the human climate niche. In press. *Proceedings of the National Academy of Sciences.*

34. The extensive literature on the evolution of cooperation and caring suggests that it can arise through several mechanisms, including the selective advantage of caring for kin and group.

An early literature on social Darwinism, which assumed that cultural traits can increase in frequency through non-genetic mechanisms, has been largely discredited as an evolutionary mechanism. Axelrod, R. and W.D. Hamilton. 1981. The evolution of cooperation. *Science* 211(4489):1390–1396. http://science.sciencemag.org/content/211/4489/1390. Nowak, M.A. 2006. Five rules for the evolution of cooperation. *Science* 314(5805):1560–1563. http://science.sciencemag.org/content/314/5805/1560.

35. Brown, K. et al. 2019. Empathy, place and identity interactions for sustainability. *Global Environmental Change* 56:11–17. https://www.sciencedirect.com/science/article/abs/pii/S0959378018307076.

36. Chapin, F.S., III et al. 2015. Earth stewardship: An initiative by the Ecological Society of America to foster engagement to sustain planet Earth. Pp. 173–194. *In* Rozzi, R. et al. (editors). *Earth Stewardship*. New York, NY: Springer.

37. Ibid.

38. The proportions of land area occupied by land managers, indigenous people, and urban residents are known only approximately, due to the difficulty of defining these groups precisely and uncertainty in the total land area that each occupies. Discussions with Erle Ellis suggest that the estimates in the text are reasonable. Ellis, E.C. and N. Ramankutty. 2008. Putting people in the map: Anthropogenic biomes of the world. *Frontiers in Ecology and the Environment* 6(8):439–447. https://doi.org/10.1890/070062.

39. Veit, P., and K. Reytar. March 20, 2017. *By the numbers: Indigenous and community land rights*. World Resources Institute. https://www.wri.org/blog/2017/03/numbers-indigenous-and-community-land-rights.

40. Ellis and Ramankutty. 2008.

41. Chapin et al. 2009. *Principles of Ecosystem Stewardship*.

42. Folke et al. 2011. Reconnecting to the biosphere.

43. Chapin et al. 2011. *Principles of Terrestrial Ecosystem Ecology*.

44. Williams, R. 1976. *Keywords: A Vocabulary of Culture and Society*. London, UK: Fontana Press.

45. Chapin et al. 2009. *Principles of Ecosystem Stewardship*.

46. Ibid.

47. Ibid.; Chapin et al. 2015. Earth stewardship.

48. Matson, P., W.C. Clark, and K. Andersson. 2016. *Pursuing Sustainability: A Guide to the Science and Practice*. Princeton, NJ: Princeton University Press.

49. Chapin et al. 2009. *Principles of Ecosystem Stewardship*; Chapin et al. 2015. Earth stewardship; Enqvist, J.P. et al. 2018. Stewardship as a boundary object for sustainability research: Linking care, knowledge and agency. *Landscape and Urban Planning* 179:17–37. https://www.sciencedirect.com/science/article/pii/S0169204618305966.

50. World Commission on Environment and Development. 1987. *Report of the World Commission on Environment and Development: Our common future*. Oxford, UK: Oxford University Press. https://sustainabledevelopment.un.org/content/documents/5987our-common-future.pdf.

51. Matson et al. 2016. *Pursuing Sustainability*.

Chapter 2

1. Holling, C.S. and G.K. Meffe. 1996. Command and control and the pathology of natural resource management. *Conservation Biology* 10(2):328–337. https://doi.org/10.1046/j.1523-1739.1996.10020328.x.

2. Easteal, S. 1981. The history of introductions of *Bufo marinus* (Amphibia: Anura); A natural experiment in evolution. *Biological Journal of the Linnean Society* 16(2):93–113. https://onlinelibrary.wiley.com/doi/abs/10.1111/j.1095-8312.1981.tb01645.x.

3. Wodzicki, K. and S. Wright. 1984. Introduced birds and mammals in New Zealand and their effect on the environment. *Tuatara* 27(2):77–104. http://nzetc.victoria.ac.nz/tm/scholarly/tei-Bio27Tuat02-t1-body-d1.html.

4. 1990: Aral Sea is "world's worst disaster." BBC News, Oct. 22, 2005. http://news.bbc.co.uk/onthisday/hi/dates/stories/october/22/newsid_3756000/3756134.stm.

5. Holling and Meffe. 1996. Command and control.

6. Chapin, F.S., III et al. (editors). 2006. *Alaska's Changing Boreal Forest*. New York, NY: Oxford University Press.

7. Schuur, E.A.G. et al. 2015. Climate change and the permafrost carbon feedback. *Nature* 520:171–179. https://www.nature.com/articles/nature14338.

8. Johnstone, J.F. et al. 2010. Changes in fire regime break the legacy lock on successional trajectories in Alaskan boreal forest. *Global Change Biology* 16(4):1281–1295. https://onlinelibrary.wiley.com/doi/abs/10.1111/j.1365-2486.2009.02051.x.

9. Mann, D.H. et al. 2012. Is Alaska's boreal forest now crossing a major ecological threshold? *Arctic, Antarctic, and Alpine Research* 44(3):319–331. https://bioone.org/journals/arctic-antarctic-and-alpine-research/volume-44/issue-3/1938-4246-44.3.319/Is-Alaskas-Boreal-Forest-Now-Crossing-a-Major-Ecological-Threshold/10.1657/1938-4246-44.3.319.full.

10. Lloyd, A.H. et al. 2006. Holocene development of the boreal forest in Alaska's changing boreal forest. Pp. 62–78. *In* Chapin, F.S., III et al. (editors). *Alaska's Changing Boreal Forest*. New York, NY: Oxford University Press.

11. Markon, C.T. et al. 2018. Alaska. Pp. 1185–1242. *In* Reidmiller, D.R. et al. (editors). *Impacts, Risks, and Adaptation in the United States: Fourth National Climate Assessment*, Vol. II. Washington DC: US Global Change Research Program. https://nca2018.globalchange.gov/chapter/26/.

12. Jenny, H. 1980. *The Soil Resource: Origin and Behavior*. New York, NY: Springer-Verlag.

13. Jenny, H. 1973. The Pygmy Forest Ecological Staircase: A description and interpretation (unpublished report available from the author at terry.chapin@alaska.edu). [Pygmy Forest Jenny 1973—Mendocino Coast Recreation].

14. DLTK. 2019. The story of Goldilocks and the three bears. https://www.dltk-teach.com/rhymes/goldilocks_story.htm.

15. Worster, D. 2004. *Dust Bowl: The Southern Plains in the 1930s*. Oxford, UK: Oxford University Press; McLeman, R.A. et al. 2014. What we learned from the Dust Bowl: Lessons in science, policy, and adaptation. *Population and Environment* 35(4):417–440. https://www.ncbi.nlm.nih.gov/pmc/articles/PMC4015056/.

16. Chapin, F.S., III, P.A. Matson, and P.M. Vitousek. 2011. *Principles of Terrestrial Ecosystem Ecology*. 2nd edition. New York, NY: Springer.

17. Land Institute. 2020. *Transforming agriculture, perennially*. https://landinstitute.org/our-work/perennial-crops/; Altieri, M.A. 1995. *Agroecology: The Science of Sustainable Agriculture*. Boulder, CO: Westview Press.

18. Barry, S., S. Larson, and M. George. 2006. California native grasslands. A historical perspective. A guide for developing realistic restoration objectives. *Grasslands*. University of California Cooperative extension. Winter 2006: 7–11. https://ucanr.edu/repository/fileaccess.cfm?article=158146&p=NNTFJW.

19. American Chestnut Foundation. *History of the American Chestnut*. https://www.acf.org/the-american-chestnut/history-american-chestnut/.

20. Leopold, A. 1949. *A Sand County Almanac*. Oxford, UK: Oxford University Press.

21. Chapin et al. 2011. *Principles of Terrestrial Ecosystem Ecology*.

22. Millar, C.I., N.L. Stephenson, and S.L. Stephens. 2007. Climate change and the forests of the future: Managing in the face of uncertainty. *Ecological Applications* 17(8):2145–2151. https://doi.org/10.1890/06-1715.1.

23. Appell, D. 2009. Can "assisted migration" save species from global warming? *Scientific American* 300: 78–80. https://www.scientificamerican.com/article/assited-migration-global-warming/.

24. Chapin et al. 2011. *Principles of Terrestrial Ecosystem Ecology*.

25. Goulson, D. et al. 2015. Bee declines driven by combined stress from parasites, pesticides, and lack of flowers. *Science* 347:1255957. https://science.sciencemag.org/content/347/6229/1255957.

26. Williams, C. 2016. These photos capture the startling effect of shrinking bee populations. *Huffington Post*. Apr. 7, https://www.huffpost.com/entry/humans-bees-china_n_570404b3e4b083f5c6092ba9.

27. Chapin et al. 2011. *Principles of Terrestrial Ecosystem Ecology*.

28. Bajaj, V., J. Ma, and S.A. Thompson. 2017. How Houston's growth created the perfect flood conditions. *New York Times*. Sept. 5, https://www.nytimes.com/interactive/2017/09/05/opinion/hurricane-harvey-flood-houston-development.html; Boburg, S. and B. Reinhard. 2017. Houston's "wild west" growth. *Washington Post*. Aug. 29, https://www.washingtonpost.com/graphics/2017/investigations/harvey-urban-planning/?utm_term=.78be55be5f9e.

29. Roll, L. and N. Bonaccorso. 2017. Rebuilt, repaired, abandoned: Five years after Sandy. *U.S. News*. Oct. 25, https://www.nbcnews.com/news/us-news/rebuilt-repaired-abandoned-five-years-after-sandy-n813696.

30. Headwaters Economics. 2019. *Solutions*. https://headwaterseconomics.org/topic/wildfire/solutions/.

31. Mathevet, R. et al. 2016. Protected areas and their surrounding territory: Socioecological systems in the context of ecological solidarity. *Ecological Applications* 26(1):5–16. https://doi.org/10.1890/14-0421.

32. Ibid.

33. Mathevet, R., F. Bousquet, and C.M. Raymond. 2018. The concept of stewardship in sustainability science and conservation biology. *Biological Conservation* 217:363–370. https://www.sciencedirect.com/science/article/pii/S000632071730407X.

34. European Union. 2016. *Life and New Partnerships for Nature Conservation*. Luxembourg: Publications Office of the European Union. https://publications.europa.eu/en/publication-detail/-/publication/c451afab-cfc6-11e5-a4b5-01aa75ed71a1/language-en.

35. Bennett, E.M. et al. 2016. Bright spots: Seeds of a good Anthropocene. *Frontiers in Ecology and the Environment* 14(8):441–448. https://doi.org/10.1002/fee.1309.

36. McKinley, D.C. et al. 2015. Investing in citizen science can improve natural resource management and environmental protection. *Issues in Ecology* 19. https://www.esa.org/wp-content/uploads/2015/09/Issue19.pdf.

37. Robbins, J. 2019. In an era of drought, Phoenix prepares for a future without Colorado River water. *Yale Environment 360*. Feb. 7, https://e360.yale.edu/features/how-phoenix-is-preparing-for-a-future-without-colorado-river-water.

Chapter 3

1. Maathai, W. 2003. *The Green Belt Movement: Sharing the Approach and the Experience.* New York, NY: Lantern Books. https://www.greenbeltmovement.org/wangari-maathai.

2. Ibid.

3. Ibid.

4. Millennium Ecosystem Assessment. 2005. *Ecosystems and Human Well-being: Synthesis.* Washington, DC: Island Press. https://www.millenniumassessment.org/documents/document.356.aspx.pdf; Daily, G.C. (editor). 1997. *Nature's Services: Societal Dependence on Natural Ecosystems.* Washington, DC: Island Press; Costanza, R. et al. 1997. The value of the world's ecosystem services and natural capital. *Nature* 387:253–260. https://www.nature.com/articles/387253a0; Chapin, F.S., III, G.P. Kofinas, and C. Folke (editors). 2009. *Principles of Ecosystem Stewardship: Resilience-Based Natural Resource Management in a Changing World.* New York, NY: Springer.

5. Millennium Ecosystem Assessment. 2005.

6. Díaz, S. et al. 2019. Pervasive human-driven decline of life on Earth points to the need for transformative change. *Science* 366, eaax3100. DOI: 10.1126/science.aaw3100. http://www.sciencemagazinedigital.org/sciencemagazine/13_december_2019/MobilePagedArticle.action?articleId=1545408&app=false#articleId1545408.

7. Chapin et al. 2009. *Principles of Ecosystem Stewardship*; Millennium Ecosystem Assessment. 2005.

8. Ibid.

9. Pope Francis. 2015. *Laudato Si'*, The Vatican. http://w2.vatican.va/content/francesco/en/encyclicals/documents/papa-francesco_20150524_enciclica-laudato-si.html.

10. Diamond, J. 2008. What's your consumption factor? *New York Times.* Jan. 2, https://www.nytimes.com/2008/01/02/opinion/02diamond.html.

11. Ellis, E.C. and N. Ramankutty. 2008. Putting people in the map: Anthropogenic biomes of the world. *Frontiers in Ecology and the Environment* 6(8):439–447. https://doi.org/10.1890/070062.

12. Roser, M., H. Ritchie, and E. Ortiz-Ospina. 2019. *World Population Growth.* May, https://ourworldindata.org/world-population-growth.

13. Barrett, S. et al. 2020. Fertility behavior and consumption patterns in the Anthropocene. *Proceedings of the National Academy of Sciences.* 117(12):6300–6307. https://www.pnas.org/content/117/12/6300.full.pdf.

14. Ibid.

15. Ritchie, H. 2017. Yields vs. land use: How the green revolution enabled us to feed a growing population. *Our World in Data.* Aug. 22, https://ourworldindata.org/yields-vs-land-use-how-has-the-world-produced-enough-food-for-a-growing-population; Ritchie, H. 2017. How much of the world's land would we need in order to feed the global population with the average diet of a given country? *Our World in Data.* Oct. 3, https://ourworldindata.org/agricultural-land-by-global-diets.

16. Matson, P.A. 2012. *Seeds of Sustainability: Lessons from the Birthplace of the Green Revolution in Agriculture.* Washington, DC: Island Press.

17. Chen, X. et al. 2014. Producing more grain with lower environmental costs. *Nature* 514:486–489. https://doi.org/10.1038/nature13609.

18. Ritchie. 2017. How much of the world's land?; Herrero, M. 2015. Livestock and the environment: What have we learned in the past decade? *Annual Review of Environment and Resources* 40:177–202. https://www.annualreviews.org/doi/abs/10.1146/annurev-environ-031113-093503.

19. Ibid.

20. Herrero. 2015. Livestock and the environment.

21. Ibid.

22. Sengupta, S. 2017. How much food do we waste? Probably more than you think. *New York Times*. Dec. 12, https://www.nytimes.com/2017/12/12/climate/food-waste-emissions.html.

23. Stefanini, S. 2019. EU set to tighten rules on palm oil for biofuels. *Climate Home News*. April 4, https://www.climatechangenews.com/2019/02/04/eu-set-tighten-rules-palm-oil-biofuels/ ; Is harvesting palm oil destroying the rainforests? December 11, 2008. *Scientific American*. https://www.scientificamerican.com/article/harvesting-palm-oil-and-rainforests/.

24. Videl, J. From the Amazon to chicken nuggets. *Guardian Weekly*. https://www.theguardian.com/guardianweekly/story/0,,1752430,00.html.

25. Ibid; Stefanini. 2019. EU set to tighten rules; Is harvesting palm oil destroying the rainforests?; Videl. From the Amazon to chicken nuggets.

26. United Nations. 2020. *REDD+ Web Platform*. https://redd.unfccc.int/fact-sheets.html.

27. Veit, P. and K. Reytar. 2017. *By the numbers: Indigenous and community land rights*. World Resources Institute. March 20, https://www.wri.org/blog/2017/03/numbers-indigenous-and-community-land-rights.

28. Doyle, A. 2012. Mangroves under threat from shrimp farms: U.N. *Reuters*. Nov. 14, https://www.reuters.com/article/us-mangroves-idUSBRE8AD1EG20121114.

29. Naylor, R.L. et al. 2000. Effect of aquaculture on world fish supplies. *Nature* 405:1017–1024. https://www.nature.com/articles/35016500.

30. Videl, J. 2017. Salmon farming in crisis: "We are seeing a chemical arms race in the seas." *The Guardian*. Apr. 1, https://www.theguardian.com/environment/2017/apr/01/is-farming-salmon-bad-for-the-environment.

31. Millennium Ecosystem Assessment. 2005.

32. Hilborn, R. and D. Ovando. 2014. Reflections on the success of traditional fisheries management. *ICES Journal of Marine Science* 71(5):1040–1046. https://academic.oup.com/icesjms/article/71/5/1040/648075; Costello, C. et al. 2016. Global fishery prospects under contrasting management regimes. *Proceedings of the National Academy of Sciences* 113(18):5125–5129. https://www.pnas.org/content/113/18/5125.

33. Millennium Ecosystem Assessment. 2005.

34. Hu, W. 2018. A billion-dollar investment in New York's water. *New York Times*. Jan. 18, https://www.nytimes.com/2018/01/18/nyregion/new-york-city-water-filtration.html.

35. Nature Conservancy. 2020. *Latin America: Creating Water Funds for People and Nature*. https://www.nature.org/en-us/about-us/where-we-work/latin-america/stories-in-latin-america/water-funds-of-south-america/.

36. Peaks to People Water Fund. 2020. *Our Solutions*. https://peakstopeople.org/our-solutions/.

37. Ricketts, T.H. et al. 2004. Economic value of tropical forest to coffee production. *Proceedings of the National Academy of Sciences* 101(34):12579–12582. https://doi.org/10.1073/pnas.0405147101.

38. When nutrient inputs to coastal estuaries are large, more algae are produced than invertebrates and fish can eat. As algae die, the resulting dead organic matter sinks to depth, where rapid decomposition, oxygen depletion, and death of fish, shrimp, and other bottom dwellers creates a dead zone of low biological activity. The dead zone that forms each spring at the mouth of the Mississippi River, for example, is about the size of the state of Connecticut. Two-thirds of the world's estuaries now have a dead zone caused by upstream nutrient pollution. Under unpolluted conditions, estuaries support some of the world's most productive fisheries, but today there is no strong fishery in the dead zones. National Oceanic and Atmospheric Administration. 2017. *Gulf of Mexico "Dead Zone" Is the Largest Ever Measured.* https://www.noaa.gov/media-release/gulf-of-mexico-dead-zone-is-largest-ever-measured.

39. Sayre, N. 2005. *Working Wilderness: The Malpai Borderlands Group and the Future of the Western Range.* Tucson, AZ: Rio Nuevo Press.

40. Ibid.

41. Jackson, L. et al. 2012. *Adaptation Strategies for Agricultural Sustainability in Yolo County, California.* Sacramento, CA: California Energy Commission. https://ww2.energy.ca.gov/2012publications/CEC-500-2012-032/CEC-500-2012-032.pdf.

42. McCoy, K. 2012. Leadership in her father's footsteps. *First Alaskans* 2012:110–112.

43. Ji, J.S. et al. 2019. Residential greenness and mortality in oldest-old women and men in China: A longitudinal study. *Lancet Planetary Health* 3(1):e17–e25. https://www.sciencedirect.com/science/article/pii/S254251961830264X?via%3Dihub.

Chapter 4

1. Millennium Ecosystem Assessment. 2005. *Ecosystems and Human Well-Being: Synthesis.* Washington, DC: Island Press. https://www.millenniumassessment.org/documents/document.356.aspx.pdf.

2. Wulf, A. 2015. *The Invention of Nature: Alexander von Humboldt's New World.* New York, NY: Alfred A. Knopf.

3. There is a huge literature that documents the causes and consequences of climate change. In 2018, scientific assessments of climate-change science were updated for the world by the Intergovernmental Panel on Climate Change (IPCC) and for the United States by the U.S. National Climate Assessment (NCA4). IPCC. 2018. Summary for policy makers. Pages 1–24 *In* Masson-Delmotte, V. et al. (editors). *Global Warming of 1.5°C above Pre-industrial Levels and Related Global Greenhouse Gas Emission Pathways, in the Context of Strengthening the Global Response to the Threat of Climate Change, Sustainable Development, and Efforts to Eradicate Poverty.* Geneva: World Meteorological Organization. https://www.ipcc.ch/sr15/chapter/spm/; Wuebbles, D.J. et al. (editors). 2017. *Climate Science Special Report: Fourth National Climate Assessment,* Vol. I. Washington, DC: US Global Change Research Program. https://science2017.globalchange.gov; US Global Change Research Program. 2018. *Impacts, Risks, and Adaptation in the United States: Fourth National Climate Assessment,* Vol. II. Washington, DC: US Global Change Research Program. https://www.globalchange.gov/browse/reports/report-brief-fourth-national-climate-assessment-volume-ii-impacts-risks-and; Davenport, C. 2018. Major climate report describes a strong risk

of crisis as early as 2040. *New York Times.* Oct. 7, https://www.nytimes.com/2018/10/07/climate/ipcc-climate-report-2040.html?action=click&module=Top%20Stories&pgtype=Homepage.

4. Chapin, F.S., III, P.A. Matson, and P.M. Vitousek. 2011. *Principles of Terrestrial Ecosystem Ecology.* 2nd edition. New York, NY: Springer; Serreze, M.C. 2010. Understanding recent climate change. *Conservation Biology* 24:10–17. https://onlinelibrary.wiley.com/doi/abs/10.1111/j.1523-1739.2009.01408.x.

5. Chapin et al. 2011. *Principles of Terrestrial Ecosystem Ecology.*

6. Lindsey, R. 2019. Climate change: Atmospheric carbon dioxide. *Climate.gov.* National Oceanic and Atmospheric Administration. Sept. 19, https://www.climate.gov/news-features/understanding-climate/climate-change-atmospheric-carbon-dioxide.

7. Ibid.

8. Global warming's six Americas. 2020. *Yale Program on Climate Change Communication.* https://climatecommunication.yale.edu/about/projects/global-warmings-six-americas/; Skeptical Science: Getting skeptical about global warming skepticism. 2020. https://www.skepticalscience.com.

9. Wuebbles et al. 2017. *Climate Science Special* Report.

10. Global warming's six Americas. 2020.

11. IPCC. 2018. *Summary for Policy Makers*; Wuebbles et al. 2017. *Climate Science Special Report*; US Global Change Research Program. 2018. *Impacts, Risks, and Adaptation*; Davenport. 2018. Major climate report.

12. Wuebbles et al. 2017. *Climate science special report.*

13. US Department of Defense. July 29, 2015. DoD releases report on security implications of climate change. *DOD News.* https://dod.defense.gov/News/Article/Article/612710/.

14. Bronen, R. and F.S. Chapin, III. 2013. Adaptive governance and institutional strategies for climate-induced community relocations in Alaska. *Proceedings of the National Academy of Sciences* 110(23):9320–9325. https://doi.org/10.1073/pnas.1210508110.

15. Smiley, D. 2017. How 4 a.m. chats persuaded Miami's Republican mayor to care about sea-level rise. *Miami Harold.* Oct. 6, https://www.miamiherald.com/news/politics-government/article177433831.html; Kolbert, E. 2015. The siege of Miami. *The New Yorker.* Dec. 14, https://www.newyorker.com/magazine/2015/12/21/the-siege-of-miami.

16. Wuebbles et al. 2017. *Climate Science Special Report*; Chapin et al. 2011. *Principles of Terrestrial Ecosystem Ecology*; Kolbert. 2015. The siege of Miami.

17. IPCC. 2018. *Summary for Policy Makers*; Wuebbles et al. 2017. *Climate Science Special Report*; US Global Change Research Program. 2018. *Impacts, Risks, and Adaptation*; Davenport. 2018. Major climate report.

18. Kolbert. 2015. The siege of Miami.

19. Wuebbles et al. 2017. *Climate Science Special Report.* The number of people vulnerable to sea-level rise is uncertain because estimates of both coastal elevation and its population density are not well known. People living within 3 feet of mean high tide are directly vulnerable to flooding, and coastal people at higher elevations are vulnerable to storm surges and disruption of infrastructure. Gillis, J. 2012. Rising sea levels seen as threat to coastal U.S. *New York Times.* March 13, https://www.nytimes.com/2012/03/14/science/earth/study-rising-sea-levels-a-risk-to-coastal-states.html.

20. IPCC. 2018. *Summary for Policy Makers*; Wuebbles et al. 2017. *Climate Science Special Report*; US Global Change Research Program. 2018. *Impacts, Risks, and Adaptation.*

21. Carrington, D. 2016. Climate change threatens ability of insurers to manage risk. *The Guardian*. Dec. 7, https://www.theguardian.com/environment/2016/dec/07/climate-change-threatens-ability-insurers-manage-risk.

22. Nelson, A. 2019. Climate change could make insurance too expensive for most people. *The Guardian*. March 21, https://www.theguardian.com/environment/2019/mar/21/climate-change-could-make-insurance-too-expensive-for-ordinary-people-report.

23. Schwartz, J. and R. Fausset. 2018. North Carolina, warned of rising seas, chose to favor development. *New York Times*. Sept. 12, https://www.nytimes.com/2018/09/12/us/north-carolina-coast-hurricane.html.

24. Bipartisan Policy Center. 2007. Impacts of global warming on North Carolina's coastal economy. June 21, https://bipartisanpolicy.org/report/impacts-global-warming-north-carolinas-coastal-economy/.

25. Ibid.

26. UK Met Office. The heatwave of 2003. https://www.metoffice.gov.uk/weather/learn-about/weather/case-studies/heatwave.

27. Mora, C. et al. 2017. Global risk of deadly heat. *Nature Climate Change* 7:501–506. https://www.nature.com/articles/nclimate3322

28. Sengupta, S. 2019. A heat wave tests Europe's defenses. Expect more. *New York Times*. July 1, https://www.nytimes.com/2019/07/01/climate/europe-heat-wave.html?action=click&module=Top%20Stories&pgtype=Homepage.

29. Pomeroy, J. 2018. As a water crisis looms in Cape Town, could it happen in Canada? *The Conversation*. Feb. 15, https://theconversation.com/as-a-water-crisis-looms-in-cape-town-could-it-happen-in-canada-90582.

30. Headwaters Economics. 2019. *Solutions*. https://headwaterseconomics.org/topic/wildfire/solutions/.

31. Minnemeyer, S., N. Harris, and O. Payne. 2017. Conserving forests could cut carbon emissions as much as getting rid of every car on earth. World Resources Institute. Nov. 27, https://www.wri.org/blog/2017/11/conserving-forests-could-cut-carbon-emissions-much-getting-rid-every-car-earth; Griscom, B.W. et al. 2017. Natural climate solutions. *Proceedings of the National Academy of Sciences* 114(44):11645–11650. http://www.pnas.org/content/114/44/11645; Bastin, J.-F. et al. 2019. The global tree restoration potential. *Science* 365(6448):76–79. https://science.sciencemag.org/content/365/6448/76 ; Carrington, D.P. July 4, 2019. Tree planting 'has mind-blowing potential' to tackle climate change. *The Guardian*. https://www.theguardian.com/environment/2019/jul/04/planting-billions-trees-best-tackle-climate-crisis-scientists-canopy-emissions?CMP=Share_iOSApp_Other.

32. Barrett, S. et al. 2014. Climate engineering reconsidered. *Nature Climate Change* 4:527–529. https://www.nature.com/articles/nclimate2278

33. Global Covenant of Mayors for Climate & Energy. https://www.globalcovenantofmayors.org/about/.

34. Hayhoe, K. 2018. The most important thing you can do to fight climate change: Talk about it. TED Talks. https://www.ted.com/talks/katharine_hayhoe_the_most_important_thing_you_can_do_to_fight_climate_change_talk_about_it?utm_campaign=tedspread&utm_medium=referral&utm_source=tedcomshare .

35. Krupnik, I. and D. Jolly (editors). 2002. *The Earth Is Faster Now*. Fairbanks, AK: ARCUS.

Chapter 5

1. Maslow, A.H. 1998. *Toward a Psychology of Being*. 3rd edition. New York, NY: John Wiley & Sons; Selva, J. 2019. Abraham Maslow, his theory and contribution to psychology. *Positive Psychology Program*. https://positivepsychologyprogram.com/abraham-maslow/.

2. Goebel, B.L. and D.R. Brown. 1981. Age differences in motivation related to Maslow's need hierarchy. *Developmental Psychology* 17(6):809–815. https://doi.org/10.1037/0012-1649.17.6.809.

3. Ibid.

4. World Bank. 2019. Poverty. https://www.worldbank.org/en/topic/poverty.

5. Bondarenko, P. 2019. Gross domestic product. *Encyclopaedia Britannica*. April 26, https://www.britannica.com/topic/gross-domestic-product; World Bank. 2019. Poverty.

6. Millennium Ecosystem Assessment. 2005. *Ecosystems and Human Well-being: Synthesis*. Washington, DC: Island Press. https://www.millenniumassessment.org/documents/document.356.aspx.pdf.

7. Stiglitz, J.E., A. Sen, and J.-P. Fitoussi. 2009. *The Measurement of Economic Performance and Social Progress Revisited*. Paris: Centre de Recherche en Économie de Sciences Po. https://www.ofce.sciences-po.fr/pdf/dtravail/WP2009-33.pdf.

8. Schreckenberg, K., G. Mace, and M. Poudyal (editors). 2018. *Ecosystem Services and Poverty Alleviation: Trade-offs and Governance*. London, UK: Routledge; Coulthard, S., J.A. McGregor, and C. White. 2018. Multiple dimensions of wellbeing in practice. Pp. 243–256. *In* Schreckenberg, K., G. Mace, and M. Poudyal (editors). *Ecosystem Services and Poverty Alleviation: Trade-offs and Governance*. London, UK: Routledge.

9. Schreckenberg et al. 2018. *Ecosystem Services and Poverty Alleviation*.

10. Schreckenberg et al. 2018. *Ecosystem Services and Poverty Alleviation*; Coulthard et al. 2018. Multiple dimensions of wellbeing; Szaboova, L. et al. 2018. Resilience and wellbeing for sustainability. Pp. 273–287. *In* Schreckenberg, K., G. Mace, and M. Poudyal (editors). *Ecosystem Services and Poverty Alleviation: Trade-offs and Governance*. London, UK: Routledge; White, S.C. 2017. Relational wellbeing: Re-centring the politics of happiness, policy and the self. *Policy & Politics* 45(2):121–136. https://www.ingentaconnect.com/content/tpp/pap/2017/00000045/00000002/art00001;jsessionid=1oghmwsq65uth.x-ic-live-02.

11. Coulthard et al. 2018. Multiple dimensions of wellbeing; White. 2017. Relational wellbeing.

12. Coulthard et al. 2018. Multiple dimensions of wellbeing.

13. United Nations. 2015. *The Millennium Development Goals Report 2015*. http://www.un.org/millenniumgoals/2015_MDG_Report/pdf/MDG%202015%20rev%20(July%201).pdf.

14. Ibid.

15. Hayhoe, K. 2018. The most important thing you can do to fight climate change: Talk about it. TED Talks. https://www.ted.com/talks/katharine_hayhoe_the_most_important_thing_you_can_do_to_fight_climate_change_talk_about_it?utm_campaign=tedspread&utm_medium=referral&utm_source=tedcomshare.

16. Brown, K. et al. 2019. Empathy, place and identity interactions for sustainability. *Global Environmental Change* 56:11–17. https://www.sciencedirect.com/science/article/abs/pii/S0959378018307076.

17. Ibid.

18. Nassauer, J.I. 2011. Care and stewardship: From home to planet. *Landscape and Urban Planning* 100(4):321–323. https://www.sciencedirect.com/science/article/pii/S0169204611000806.

19. Statista. 2020. Median household income in the United States in 2018, by educational attainment of householder. https://www.statista.com/statistics/233301/median-household-income-in-the-united-states-by-education/.

20. Duckworth, A.L. et al. 2007. Grit: Perseverance and passion for long-term goals. *Journal of Personality and Social Psychology* 92(6):1087–1101. https://psycnet.apa.org/record/2007-07951-009.

21. Folke, C. et al. 2011. Reconnecting to the biosphere. *Ambio* 40(7):719–738. https://www.ncbi.nlm.nih.gov/pmc/articles/PMC3357749/.

22. Millennium Ecosystem Assessment. 2005. *Ecosystems and Human Well-being: Health Synthesis*. Washington, DC: Island Press. https://www.millenniumassessment.org/documents/document.357.aspx.pdf.

23. Szaboova. et al. 2018. Resilience and wellbeing for sustainability; Millennium Ecosystem Assessment. 2005. *Ecosystems and Human Well-being*.

24. Szaboova. et al. 2018. Resilience and wellbeing for sustainability.

25. To do with the price of fish. May 10, 2007. *The Economist*. https://www.economist.com/finance-and-economics/2007/05/10/to-do-with-the-price-of-fish.

26. Winsor, M. 2015. African cell phone use: Sub-Saharan Africa sees surge in mobile ownership, study finds. *International Business Times*. April 15, http://www.ibtimes.com/african-cell-phone-use-sub-saharan-africa-sees-surge-mobile-ownership-study-finds-1883449.

27. Louv, R. 2008. *Last Child in the Woods: Saving Our Children from Nature-Deficit Disorder*. Chapel Hill, NC: Algonquin Books.

Chapter 6

1. Chapin, F.S., III and C.N. Knapp. 2015. Sense of place: A process for identifying and negotiating potentially contested visions of sustainability. *Environmental Science and Policy* 53:38–46. https://doi.org/10.1016/j.envsci.2015.04.012; Fresque-Baxter, J.A. and D. Armitage. 2012. Place identity and climate change adaptation: A synthesis and framework for understanding. *Climate Change* 3(3):251–266. https://doi.org/10.1002/wcc.164; Soga, M. and K.J. Gaston. 2016. Extinction of experience: The loss of human-nature interactions. *Frontiers in Ecology and the Environment* 14(2):94–101. https://esajournals.onlinelibrary.wiley.com/doi/full/10.1002/fee.1225; Louv, R. 2008. *Last Child in the Woods: Saving Our Children from Nature-Deficit Disorder*. Chapel Hill, NC: Algonquin Books .

2. Chapin and Knapp. 2015. Sense of place; Brown, K. et al. 2019. Empathy, place and identity interactions for sustainability. *Global Environmental Change* 56:11–17. https://www.sciencedirect.com/science/article/abs/pii/S0959378018307076; Enqvist, J.P. et al. 2018. Stewardship as a boundary object for sustainability research: Linking care, knowledge and agency. *Landscape and Urban Planning* 179:17–37. https://www.sciencedirect.com/science/article/pii/S0169204618305966.

3. Chapin and Knapp. 2015. Sense of place; Fresque-Baxter and Armitage. 2012. Place identity and climate change adaptation; Soga and Gaston. 2016. Extinction of experience.

4. Brown et al. 2019. Empathy, place and identity interactions for sustainability; Enqvist et al. 2018. Stewardship as a boundary object.

5. Ibid.

6. Ibid.

7. WaterFire Providence. 2020. https://waterfire.org/about/.

8. Ibid.

9. Gillis, J. and H. Harvey. 2018. Cars are ruining our cities. *New York Times*. April 25, https://www.nytimes.com/2018/04/25/opinion/cars-ruining-cities.html.

10. Greening of Detroit. 2019. https://www.greeningofdetroit.com.

11. Trust for Public Land. 2019. https://www.tpl.org/10minutewalk.

12. Eden Place Nature Center. http://www.edenplacenaturecenter.org.

13. Ibid.

14. http://www.edenplacenaturecenter.org/urbaane.html.

15. Pizza garden seed collection. *Interfaith Power & Light*. https://interfaith-power-light.myshopify.com/products/pizza-garden-seed-collection.

16. Soga and Gaston. 2016. Extinction of experience; Louv. 2008. *Last Child in the Woods*.

17. Nassauer, J.I. 2011. Care and stewardship: From home to planet. *Landscape and UrbanPlanning* 100(4): 321–323. https://www.sciencedirect.com/science/article/pii/S0169204611000806.

18. Wilson, E.O. 1984. *Biophilia*. Boston, MA: Harvard University Press.

19. Soga, M. and K.J. Gaston. 2018. Shifting baseline syndrome: Causes, consequences, and implications. *Frontiers in Ecology and the Environment* 16:222–230 https://esajournals.onlinelibrary.wiley.com/doi/full/10.1002/fee.1794.

20. Nassauer. 2011. Care and stewardship.

21. Chapin, F.S., III, G.P. Kofinas, and C. Folke (editors). 2009. *Principles of Ecosystem Stewardship: Resilience-Based Natural Resource Management in a Changing World*. New York, NY: Springer; panels from Steffen, W. et al. (editors). 2004. *Global Change and the Earth System: A Planet Under Pressure*. New York, NY: Springer-Verlag.

22. Diamond, J. 2008. What's your consumption factor? *New York Times*. Jan. 2, https://www.nytimes.com/2008/01/02/opinion/02diamond.html; Story of Stuff Project. https://storyofstuff.org/movies/story-of-stuff/; Ecological footprint. *Global Footprint Network*. https://www.footprintnetwork.org/our-work/ecological-footprint/.

23. Diamond. 2008. What's your consumption factor?

24. Ecological footprint. *Global Footprint Network*; Friedman, M. 1957. *A Theory of the Consumption Function*. Princeton, NJ: Princeton University Press.

25. Ecological footprint. *Global Footprint Network*.

26. Amadeo, K. 2019. Income inequality in America. *The Balance*. Dec. 16, https://www.thebalance.com/income-inequality-in-america-3306190; Reinicke, C. 2018. US income inequality continues to grow. *CNBC*. July 19, https://www.cnbc.com/2018/07/19/income-inequality-continues-to-grow-in-the-united-states.html.

27. Income inequality. 2015. *OECD Data*. https://data.oecd.org/inequality/income-inequality.htm.

28. Zimmerman, J. 2015. 6 in 10 Americans will experience poverty. *Financial Times*. July 24, http://time.com/money/3966403/six-in-ten-americans-will-experience-poverty/.

29. Meadow, A.M. 2012. Assessing access to local food system initiatives in Fairbanks Alaska. *Journal of Agriculture, Food Systems, and Community Development* 2(2):6. https://doi.org/10.5304/jafscd.2012.022.006.

30. Easterlin, R.A. et al. 2010. The happiness-income paradox revisited. *Proceedings of the National Academy of Sciences* 107(52):22463–22468. https://doi.org/10.1073/pnas.1015962107; Diener, E. and M.E.P. Seligman. 2004. Beyond money: Toward an economy of well-being. *Psychological Science in the Public Interest* 5:1–31. https://journals.sagepub.com/doi/full/10.1111/j.0963-7214.2004.00501001.x.

31. Ibid.

32. Diener and Seligman. 2004. Beyond money.

33. Hadhazy, A. 2016. Here's the truth about the "planned obsolescence" of tech. *BBC*. June 12, http://www.bbc.com/future/story/20160612-heres-the-truth-about-the-planned-obsolescence-of-tech.

34. Corkery, M. and S. Cowley. 2017. Household debt makes a comeback in the U.S. *New York Times*. May 17, https://www.nytimes.com/2017/05/17/business/dealbook/household-debt-united-states.html.

35. Sengupta, S. 2017. How much food do we waste? Probably more than you think. *New York Times*. Dec. 12, https://www.nytimes.com/2017/12/12/climate/food-waste-emissions.html; Hingle, M. 2015. Next time you're searching for something to eat, shop your refrigerator first. *The Hill*. Nov. 20, https://thehill.com/blogs/congress-blog/260825-next-time-youre-searching-for-something-to-eat-shop-your-refrigerator.

36. McMahon, J. 2017. The four most effective things you can do about climate change, according to science. *Forbes*. July 13, https://www.forbes.com/sites/jeffmcmahon/2017/07/13/the-four-most-effective-things-you-can-do-about-climate-change-according-to-science/#5905df9d635c; Wynes, S. and K.A. Nicholas. 2017. The climate mitigation gap: Education and government recommendations miss the most effective individual actions. *Environmental Research Letters* 12:074024. https://iopscience.iop.org/article/10.1088/1748-9326/aa7541.

37. Carbon footprint calculator. *US Environmental Protection Agency*. https://www3.epa.gov/carbon-footprint-calculator/; What you can do about climate change. *US Environmental Protection Agency*. https://19january2017snapshot.epa.gov/climatechange/what-you-can-do-about-climate-change_.html; Albeck-Ripka, L. How to reduce your carbon footprint. *New York Times*. https://www.nytimes.com/guides/year-of-living-better/how-to-reduce-your-carbon-footprint?campaignId=74XU8; Hawkin, P. (editor). 2017. *Drawdown: The Most Comprehensive Plan Ever Proposed to Reverse Global Warming*. New York, NY: Penguin Books; Kalmas, P. 2017. *Being the Change. Live Well and Spark a Climate Revolution*. Gabriola Island, Canada: New Society Publishers.

38. Kalmas. 2017. *Being the Change*.

39. Goldstein, N.J., R.B. Cialdini, and V. Griskevicius. 2008. A room with a viewpoint: Using social norms to motivate environmental conservation in hotels. *Journal of Consumer Research* 35(3):472–482. https://doi.org/10.1086/586910.

40. Ibid.

41. Bohner, G. and L.E. Schlüter. 2014. A room with a viewpoint revisited: Descriptive norms and hotel guests' towel reuse behavior. *PLoS ONE* 9(8):e104086. https://doi.org/10.1371/journal.pone.0104086.

42. Cialdini, R.B. and N.J. Goldstein. 2004. Social influence: Compliance and conformity. *Annual Review of Psychology* 55:591–621. https://www.annualreviews.org/doi/abs/10.1146/annurev.psych.55.090902.142015; Nyborg, K. et al. 2016. Social norms as solutions. *Science* 354:42–43. http://science.sciencemag.org/content/354/6308/42.

43. Nyborg, K. et al. 2016. Social norms as solutions.

44. Ibid.

45. Nassauer. 2011. Care and stewardship..

Chapter 7

1. Chapin, F.S., III et al. 2016. Community-empowered adaptation for self-reliance. *Current Opinion in Environmental Sustainability* 19:67–75. https://doi.org/10.1016/j.cosust.2015.12.008.

2. Ibid.

3. Danielsen, F. et al. 2010. Environmental monitoring: The scale and speed of implementation varies according to the degree of people's involvement. *Journal of Applied Ecology* 47(6):1166–1168. https://doi.org/10.1111/j.1365-2664.2010.01874.x.

4. Hahn, T. et al. 2006. Trust-building, knowledge generation and organizational innovations: The role of a bridging organization for adaptive comanagement of a wetland landscape around Kristianstad, Sweden. *Human Ecology* 34(4):573–592. https://link.springer.com/article/10.1007/s10745-006-9035-z.

5. Danielsen et al. 2010. Environmental monitoring.

6. Paulson, A. 2018. Why these young Republicans see hope in climate action. *Christian Science Monitor*. June 28, https://www.csmonitor.com/Environment/2018/0628/Why-these-young-Republicans-see-hope-in-climate-action?utm_source=EHN&utm_campaign=61030e36d9-RSS_EMAIL_CAMPAIGN&utm_medium=email&utm_term=0_8573f35474-61030e36d9-99404265.

7. Students for Carbon Dividends. 2019. https://www.s4cd.org.

8. Young Conservatives for Energy Reform. 2018. http://yc4er.org.

9. Paulson. 2018. Why these young Republicans see hope.

10. Peterson, G.D. et al. 2003. Assessing future ecosystem services: A case study of the Northern Highlands Lake District, Wisconsin. *Conservation Ecology* 7(3):1. http://www.consecol.org/vol7/iss3/art1/.

11. Carpenter, S.R. et al. 2015. Plausible futures of a social-ecological system: Yahara watershed, Wisconsin, USA. *Ecology and Society* 20(2):10. http://dx.doi.org/10.5751/ES-07433-200210.

12. Kahneman, D. 2012. *Thinking, Fast and Slow*. New York, NY: Penguin Press.

13. Ibid.

14. Furth-Matzkin, M. and C.R. Sunstein. 2018. Social influences on policy preferences: Conformity and reactance. *Minnesota Law Review* 101:1339–1379. http://www.minnesotalawreview.org/wp-content/uploads/2018/02/Sunstein_MLR.pdf.

15. Ibid.

16. Oreskes, N. and E.M. Conway. 2010. *Merchants of Doubt: How a Handful of Scientists Obscured the Truth on Issues from Tobacco Smoke to Global Warming*. London, UK: Bloomsbury Press.

17. Ibid.

18. Kasprak, A. 2016. Did 30,000 scientists declare climate change a hoax? *Snopes*. Oct. 24, https://www.snopes.com/fact-check/30000-scientists-reject-climate-change/.

19. Arnold, E. 2018. Doom and gloom: The role of the media in public disengagement on climate change. *Shorenstein Center on Media, Politics and Public Policy*. May 29, https://shorensteincenter.org/media-disengagement-climate-change/.

20. Baumeister, R.F. et al. 2001. Bad is stronger than good. *Review of General Psychology* 5:323–370. https://journals.sagepub.com/doi/abs/10.1037/1089-2680.5.4.323.

21. Global warming's six Americas. 2016. Yale Program on Climate Change Communication. https://climatecommunication.yale.edu/about/projects/global-warmings-six-americas/.

22. Tanana Chiefs Conference. https://www.tananachiefs.org/yritfc/.

23. Diener, E. 2000. Subjective well-being: The science of happiness and a proposal for a national index. *American Psychologist* 55(1):34–43. https://psycnet.apa.org/buy/2000-13324-004; Fredrickson, B.L. and C. Branigan. 2005. Positive emotions broaden the scope of attention and thought-action repertoires. *Cognition and Emotion* 19(3):313–332. https://doi.org/10.1080/02699930441000238; Seligman, M.E.P. 2018. *The Hope Circuit. A Psychologist's Journey from Helplessness to Optimism.* New York, NY: Hachette Book Group.

24. Danielsen et al. 2010. Environmental monitoring.

25. Cash, D.W. et al. 2003. Knowledge systems for sustainable development. *Proceedings of the National Academy of Sciences* 100(14):8086–8091. https://doi.org/10.1073/pnas.1231332100.

26. Lejano, R.P., J. Tavares-Reager, and F. Berkes. 2013. Climate and narrative: Environmental knowledge in everyday life. *Environmental Science and Policy* 31:61–70. https://doi.org/10.1016/j.envsci.2013.02.009; Cruikshank, J. 1998. *The Social Life of Stories: Narrative and Knowledge in the Yukon* Territory. Lincoln, NE: University of Nebraska Press.

27. De Witt, A. 2016. Understanding our polarized political landscape requires a long deep look at our worldviews. *Scientific American*. June 28, https://blogs.scientificamerican.com/guest-blog/understanding-our-polarized-political-landscape-requires-a-long-deep-look-at-our-worldviews/.

28. Seligman. 2018. *The Hope Circuit.*

29. When agreement on worldviews or overriding principles is unlikely, it may still be possible to achieve "incompletely theorized agreement" on particular outcomes that participants want but are not explicitly tied to incompatible worldviews. Sunstein, C.R. 1995. Incompletely theorized agreements. *Harvard Law Review* 108(7):1733–1772. https://www.jstor.org/stable/1341816?seq=1#page_scan_tab_contents.

30. Hayhoe, K. 2018. The most important thing you can do to fight climate change: Talk about it. TED Talks. https://www.ted.com/talks/katharine_hayhoe_the_most_important_thing_you_can_do_to_fight_climate_change_talk_about_it?utm_campaign=tedspread&utm_medium=referral&utm_source=tedcomshare.

31. Ibid.

32. Ibid.

33. Hall, T. 2019. *Writing to Persuade. How to Bring People over to Your Side.* New York, NY. Liveright Publishing Corporation; Here and Now. July 2, 2019. Former New York Times on writing to get someone's attention—and maybe change their mind. https://www.wbur.org/hereandnow/2019/07/02/trish-hall-writing-to-persuade-book.

34. Evangelical declaration on the care of Creation. 2018. *Evangelical Environmental Network.* https://creationcare.org/what-we-do/an-evangelical-declaration-on-the-care-of-creation.html.

35. Hall. 2019. *Writing to Persuade.*

36. Ray, M. 2014. Trans-Alaska Pipeline. *Encyclopaedia Britannica.* May 16, https://www.britannica.com/topic/Trans-Alaska-Pipeline.

Chapter 8

1. Leverage points for sustainability transformation. Leuphana University. https://leveragepoints.org/project-overview/.
2. University of Alaska Fairbanks. 2016. Sharing, cooperation key to Arctic villages. *ScienceDaily*. Nov. 23, https://www.sciencedaily.com/releases/2016/11/161123141845. htm; BurnSilver, S. et al. 2016. Are mixed economies persistent or transitional? Evidence using social networks from arctic Alaska. *American Anthropologist* 118(1):121–129. https://doi.org/10.1111/aman.12447.
3. Lawson, V. 1997. Geographies of care and responsibility. *Annals of the Association of American Geographers* 97(1):1–11. https://doi.org/10.1111/j.1467-8306.2007.00520.x; Brown, K. et al. 2019. Empathy, place and identity interactions for sustainability. *Global Environmental Change* 56:11–17. https://www.sciencedirect.com/science/article/abs/pii/S0959378018307076.
4. Lawson. 1997. Geographies of care and responsibility.
5. Muskoka Summit on the Environment. Feb. 5, 2018. How can we restore our relationship with nature? https://muskokasummit.org/blog/how-can-we-restore-our-relationship-with-nature/.
6. Wondolleck, J.M. and S.L. Yaffee. 2000. *Making Collaboration Work: Lessons from Innovation in Natural Resource Management*. Washington, DC: Island Press.
7. Ansell, C. and A. Gash. 2007. Collaborative governance in theory and practice. *Journal of Public Administration Research and Theory* 18:543–571. https://academic.oup.com/jpart/article-abstract/18/4/543/1090370.
8. Mothers Against Drunk Driving. 2019. https://www.madd.org/about-us/our-story/.
9. Nyborg, K. et al. 2016. Social norms as solutions. *Science* 354:42-43. http://science. sciencemag.org/content/354/6308/42.
10. Rittel, H.W.J. and M.M. Webber. 1973. Dilemmas in a general theory of planning. *Policy Sciences* 4(2):155–169. https://link.springer.com/article/10.1007/BF01405730; Chapin, F.S., III et al. 2008. Increasing wildfire in Alaska's boreal forest: Pathways to potential solutions of a wicked problem. *BioScience* 58(6):531–540. https://academic.oup.com/bioscience/article/58/6/531/235993.
11. Hawkin, P. 2007. *How the Largest Movement in the World Came into Being, and Why No One Saw It Coming*. New York, NY: Viking Press.
12. Foderaro, L.W. 2014. Taking a call for climate change to the streets. *New York Times*. Sept. 21, https://www.nytimes.com/2014/09/22/nyregion/new-york-city-climate-change-march. html?_r=0.
13. Fairbanks Climate Action Coalition. http://fairbanksclimateaction.org/history-of-fcac/.
14. Tallis, H. et al. 2017. *Bridge collaborative practitioner's guide: Principles and guidance for cross-sector action planning and evidence evaluation*. Washington, DC: Nature Conservancy. http://bridgecollaborativeglobal.org/wp-content/uploads/2018/02/Practitioners_Guide_Final_2.pdf.
15. Women's March. 2017. Jan. 1, https://www.history.com/this-day-in-history/womens-march.
16. Biodiversity Funders Group. https://biodiversityfunders.org.
17. American Wind Wildlife Institute. https://awwi.org/about-us/history/#section-the-genesis-of-awwi.

18. Ibid.
19. Wondolleck and Yaffee. 2000. *Making Collaboration Work.*
20. Arctic Council Sept. 13, 2018. The Arctic Council: A backgrounder. https://www.arctic-council.org/index.php/en/about-us.
21. Landscape Conservation Cooperative Network. https://lccnetwork.org.
22. Blackfoot Challenge: Better communities through cooperative conservation. 2019. www.blackfootchallenge.org.
23. Tallis, H. et al. 2017. *Bridge collaborative practitioner's guide*; Hahn et al. 2006. Trust-building, knowledge generation and organizational innovations.
24. Wondolleck and Yaffee. 2000. *Making Collaboration Work.*
25. Holling, C.S. and G.K. Meffe. 1996. Command and control and the pathology of natural resource management. *Conservation Biology* 10(2):328–337. https://doi.org/10.1046/j.1523-1739.1996.10020328.x.
26. Wondolleck and Yaffee. 2000. *Making Collaboration Work*; Ansell and Gash. 2007. Collaborative governance.
27. Ibid.
28. Wondolleck and Yaffee. 2000. *Making Collaboration Work.*
29. Ibid.
30. Ibid; Ansell and Gash. 2007. Collaborative governance.
31. Sunstein, C.R. 1995. Incompletely theorized agreements. *Harvard Law Review* 108(7):1733–1772. https://www.jstor.org/stable/1341816?seq=1#page_scan_tab_contents.
32. Goldenberg, S. Feb. 14, 2012. Leak exposes how Heartland Institute works to undermine climate science. *The Guardian.* https://www.theguardian.com/environment/2012/feb/15/leak-exposes-heartland-institute-climate.
33. The Aspen Institute. https://www.aspeninstitute.org.
34. Goldenberg. 2012. Leak exposes how Heartland Institute works
35. Cuyahoga River fire. *Ohio History Connection.* http://www.ohiohistorycentral.org/w/Cuyahoga_River_Fire; Grant, J. 2017. How a burning river helped create the Clean Water Act. *Allegheny Front.* April 21, https://www.alleghenyfront.org/how-a-burning-river-helped-create-the-clean-water-act/.
36. What is the polluter pays principle? 2018. *London School of Economics and Political Science, Grantham Research Institute on Climate Change and the Environment.* May 11, http://www.lse.ac.uk/GranthamInstitute/faqs/what-is-the-polluter-pays-principle/.
37. Ibid.
38. World Bank. 2018. Pricing Carbon. http://www.worldbank.org/en/programs/pricing-carbon; Kalmas, P. 2017. The trick to make capitalism help solve climate change. *Yes Magazine.* July 12, https://www.yesmagazine.org/planet/the-trick-to-make-capitalism-help-solve-climate-change-20170712; What is emissions trading? July 5, 2011. *The Guardian.* https://www.theguardian.com/environment/2011/jul/05/what-is-emissions-trading.
39. World Bank. 2018. Pricing Carbon; Kalmas. 2017. The trick to make capitalism help solve climate change.
40. Kalmas. 2017. The trick to make capitalism help solve climate change.
41. Students for Carbon Dividends. 2019. https://www.s4cd.org; Bartsch, K., Aug. 1, 2019. Are conservatives embracing a carbon tax? *National Review.* https://www.nationalreview.com/2019/08/are-conservatives-embracing-carbon-tax/.

42. What is emissions trading? 2011; World Bank. 2018. Pricing Carbon.

43. Waring, T. and J. Acheson. 2018. Evidence of cultural group selection in territorial lob-stering in Maine. *Sustainability Science* 13(1):21–34. https://link.springer.com/article/10.1007/s11625-017-0501-x.

44. What is ecolabelling? 2019. *Global Ecolabelling Network.* https://globalecolabelling.net/what-is-eco-labelling/.

45. LEED rating system: Green building leadership is LEED. *US Green Building Council.* https://new.usgbc.org/leed; Assessing the sustainability of a product or building is not easy. Life-cycle assessment enables people to estimate the environmental impact of pro-ducing and using a product. Cradle-to-grave life-cycle assessment, for example, considers the environmental impacts of extracting the raw materials, manufacturing the product, using the product, and disposing of or recycling the product at the end of its useful life. Most assessments of product sustainability are, however, less complete. Sartori, I. and A.G. Hestnes. 2007. Energy use in the life cycle of conventional and low-energy buildings: A re-view article. *Energy and Buildings* 39(3):249–257. https://www.sintef.no/globalassets/up-load/energi/transes/article_life-cyle-energy_enb.pdf.

46. Howard, E. June 23, 2015. A beginner's guide to fossil fuel divestment. *The Guardian.* https://www.theguardian.com/environment/2015/jun/23/a-beginners-guide-to-fossil-fuel-divestment.

47. Howard. 2015. A beginner's guide; Neslen, A. Oct. 3, 2017. Catholic church to make record divestment from fossil fuels. *The Guardian* https://www.theguardian.com/environment/2017/oct/03/catholic-church-to-make-record-divestment-from-fossil-fuels.

48. Williams, A., G. Whiteman, and J.N. Parker. 2019. Backstage interorganizational collabora-tion: Corporate endorsement for the sustainable development goals. *Academy of Management Discoveries* 5(4):367–395. *https://journals.aom.org/doi/10.5465/amd.2018.0154*; World Business Council for Sustainable Development. https://www.wbcsd.org.

49. Unilever. https://www.unilever.com/about/.

50. Mark, A.F. 2015. *Standing My Ground: A Voice for Nature Conservation.* Dunedin, NZ: Otago University Press.

51. Chapin, F.S., III et al. 2012. Design principles for social-ecological transformation toward sustainability: Lessons from New Zealand sense of place. *Ecosphere* 3:40. http://dx.doi.org/10.1890/ES12-00009.1.

Chapter 9

1. ABR, Inc. https://www.abrinc.com/about/.

2. Melillo, J.M., T.C. Richmond, and G.W. Yohe (editors). 2014. *Climate Change Impacts in the United States: The Third National Climate Assessment.* US Global Change Research Program. https://nca2014.globalchange.gov.

3. IPCC. 2018. Summary for policy makers. *In* Masson-Delmotte, V. et al. (editors). *Global Warming of 1.5°C above Pre-industrial Levels and Related Global Greenhouse Gas Emission Pathways, in the Context of Strengthening the Global Response to the Threat of Climate Change, Sustainable Development, and Efforts to Eradicate Poverty.* Geneva: World Meteorological Organization. https://www.ipcc.ch/sr15/chapter/summary-for-policy-makers/.

4. Griffith, B. et al. 2002. The Porcupine caribou herd. Pages 8–37 *In* Douglas, D.C., P.E. Reynolds, and E.B. Rhode (editors). *Arctic Refuge Coastal Plain Terrestrial Wildlife Research Summaries.* Washington, DC: Geological Survey (U.S.). https://alaska.usgs.gov/products/pubs/2002/2002-USGS-BRD-BSR-2002-0001.pdf; U.S. Fish and Wildlife Service. https://www.worldcat.org/title/arctic-refuge-coastal-plain-terrestrial-wildlife-research-summaries/oclc/50010055.

5. Della Porta, D. and M. Diani. 2006. *Social Movements: An Introduction.* 2nd edition. Oxford, UK: Blackwell.

6. League of Women Voters. https://www.lwv.org/about-us/history.

7. Fairbanks Climate Action Coalition. http://fairbanksclimateaction.org/history-of-fcac/.

8. Adaptation Clearinghouse. 2011. Alaska's climate change strategy: Addressing impacts in Alaska. https://www.adaptationclearinghouse.org/resources/alaska-s-climate-change-strategy-addressing-impacts-in-alaska.html.

9. Baker, S. and F.S. Chapin, III. 2018. Going beyond "it depends": The role of context in shaping participation in natural resource management. *Ecology and Society* 23(1):20. https://doi.org/10.5751/ES-09868-230120; Van Tatenhove, J.P.M. and P. Leroy. 2003. Environment and participation in a context of political modernisation. *Environmental Values* 12(2):155–174. https://doi.org/10.3197/096327103129341270.

10. EPA. How citizens can comment and participate in the National Environmental Policy Act process. https://www.epa.gov/nepa/how-citizens-can-comment-and-participate-national-environmental-policy-act-process

11. Baker and Chapin 2018. Going beyond "it depends."

12. Adaptation Clearinghouse. 2011. Alaska's climate change strategy

13. Bronen, R. and F.S. Chapin, III. 2013. Adaptive governance and institutional strategies for climate-induced community relocations in Alaska. *Proceedings of the National Academy of Sciences* 110(23):9320–9325. https://doi.org/10.1073/pnas.1210508110.

14. Kolstad, C.D. March, 2017. *What is killing the US coal industry?* Stanford Institute for Economic Policy Research. https://siepr.stanford.edu/research/publications/what-killing-us-coal-industry. Box 9.1 is based on this reference.

15. Nature Conservancy. Oct. 15, 2018. *The science of sustainability.* https://www.nature.org/en-us/what-we-do/our-insights/perspectives/the-science-of-sustainability/.

16. Hawkins, N.J. et al. 2019. The evolutionary origins of pesticide resistance. *Biological Reviews Cambridge Philosophical Society* 94(1):135–155. https://www.pbs.org/wgbh/evolution/library/10/1/l_101_02.html

17. Land Institute. 2020. *Transforming agriculture, perennially.* https://landinstitute.org/our-work/perennial-crops/; Altieri, M.A. 1995. *Agroecology: The Science of Sustainable Agriculture.* Boulder, CO: Westview Press.

18. Ehle, J.M. 1965. *The Free Men.* New York, NY: Harper & Row.

19. Ibid.

20. The way it happened: Chester in perspective. Nov. 16, 1963. *Swarthmore College Phoenix* 84(11):1,4.

21. Satell, G. and S. Popovic. Jan. 27, 2017. How protests become successful social movements. *Harvard Business Review.* https://hbr.org/2017/01/how-protests-become-successful-social-movements.

22. Satell and Popovic. 2017. How protests become successful social movements; Mazumder, S. 2018. The persistent effect of the U.S. civil rights protests on political attitudes. *American*

Journal of Political Science 62(4):922–935. https://onlinelibrary.wiley.com/doi/full/10.1111/ajps.12384.

23. Mazumder. 2018. The persistent effect of the U.S. civil rights protests.

24. Gillion, D.Q. 2013. *The Political Power of Protest: Minority Activism and Shifts in Public Policy.* Cambridge, UK: Cambridge University Press.

25. Resilience Alliance. https://www.resalliance.org/background.

26. Folke, C. et al. 2011. Reconnecting to the biosphere. *Ambio* 40(7):719–738. https://www.ncbi.nlm.nih.gov/pmc/articles/PMC3357749/.

27. Resilience Alliance. https://www.resalliance.org/background; Olsson, P., C. Folke, and T.P. Hughes. 2008. Navigating the transition to ecosystem-based management of the Great Barrier Reef, Australia. *Proceedings of the National Academy of Sciences* 105(28):9489–9494. https://doi.org/10.1073/pnas.0706905105.

28. Olsson et al. 2008. Navigating the transition; Chapin, F.S., III et al. 2010. Ecosystem stewardship: Sustainability strategies for a rapidly changing planet. *Trends in Ecology and Evolution* 25(4):241–249. https://www.sciencedirect.com/science/article/pii/S0169534709003255.

29. Ibid.

30. Grameen Bank Facts. *The Nobel Prize.* https://www.nobelprize.org/prizes/peace/2006/grameen/facts/.

31. Olsson et al. 2008. Navigating the transition; Hughes, T.P. et al. 2017. Global warming and recurrent mass bleaching of corals. *Nature* 543:373–377. https://www.nature.com/articles/nature21707.

32. Harden, G. 1968. The tragedy of the commons. *Science* 162(3859):1243–1248. https://science.sciencemag.org/content/162/3859/1243.full.

33. Ostrom, E., M.A. Janssen, and J.M. Anderies. 2007. Going beyond panaceas. *Proceedings of the National Academy of Sciences* 104(39):15176–15178. http://www.pnas.org/content/104/39/15176.full; Ostrom, E. 1990. *Governing the Commons: The Evolution of Institutions for Collective Action.* Cambridge, UK: Cambridge University Press.

Chapter 10

1. Wuebbles et al. 2017. *Climate Science Special Report*; IPCC. 2018. Summary for policy makers. *In* Masson-Delmotte, V. et al. (editors). *Global Warming of 1.5°C above Pre-industrial Levels and Related Global Greenhouse Gas Emission Pathways, in the Context of Strengthening the Global Response to the Threat of Climate Change, Sustainable Development, and Efforts to Eradicate Poverty.* Geneva: World Meteorological Organization. https://www.ipcc.ch/sr15/chapter/summary-for-policy-makers/; US Global Change Research Program. 2018. *Impacts, Risks, and Adaptation in the United States: Fourth National Climate Assessment*, Vol. II. Washington, DC: US Global Change Research Program. https://www.globalchange.gov/browse/reports/report-brief-fourth-national-climate-assessment-volume-ii-impacts-risks-and.

2. Ibid.

3. Xu, C. et al. In press. Future of the human climate niche. In press. *Proceedings of the National Academy of Sciences.*

4. Steffen, W. et al. 2018. Trajectories of the earth system in the Anthropocene. *Proceedings of the National Academy of Sciences* 115(33):8252–8259. https://doi.org/10.1073/pnas.1810141115.

5. Wuebbles et al. 2017. *Climate Science Special Report*; IPCC. 2018. Summary for policy makers; US Global Change Research Program. 2018. *Impacts, Risks, and Adaptation*; Xu, C. et al. In press. Future of the human climate niche; Steffen et al. 2018. Trajectories of the earth system.

6. O'Brien, K. 2018. Taking climate change seriously: From adaptation to transformation. Youtube video of a seminar presented at the Stockholm Resilience Center. https://www.youtube.com/watch?v=Znu1IfxvBKc; O'Brien, K. and L. Sygna. 2013. Responding to climate change: The three spheres of transformation. *Proceedings of Transformation in a Changing Climate*. Oslo, Norway. University of Oslo. This paper was published by the non-profit organization cCHANGE. https://www.sv.uio.no/iss/english/research/projects/adaptation/publications/1-responding-to-climate-change---three-spheres-of-transformation_obrien-and-sygna_webversion_final.pdf.

7. Clark, J. What does the 'pursuit of happiness' mean in the Declaration of Independence? Howstuffworks. https://science.howstuffworks.com/life/inside-the-mind/emotions/pursuit-of-happiness-meaning.htm.

8. Kelly, A. 2012. Gross national happiness in Bhutan: The big idea from a tiny state that could change the world. *The Guardian*. Dec. 1, https://www.theguardian.com/world/2012/dec/01/bhutan-wealth-happiness-counts.

9. Ibid.

10. Kubiszewski, I. et al. 2013. Beyond GDP: Measuring and achieving global genuine progress. *Ecological Economics* 93:57–68. https://www.sciencedirect.com/science/article/pii/S0921800913001584?via%3Dihub; Genuine Progress Indicator. *Gross National Happiness USA*. https://gnhusa.org/genuine-progress-indicator/.

11. Klein, N. 2014. *This Changes Everything: Capitalism vs the Climate*. New York, NY: Simon & Schuster.

12. Friedman, L.S. 2002. *The Microeconomics of Public Policy Analysis*. Princeton: Princeton University Press..

13. Guerry, A.D. et al. 2015. Natural capital and ecosystem services informing decisions: From promise to practice. *Proceedings of the National Academy of Sciences* 112(24):7348–7355. https://www.pnas.org/content/112/24/7348.

14. Ibid.

15. DeSilver, D. 2017. 5 facts about the minimum wage. *Pew Research Center*. https://www.pewresearch.org/fact-tank/2017/01/04/5-facts-about-the-minimum-wage/.

16. State & federal minimum wage rates. 2020. *Labor Law Center*. Jan. 1, https://www.laborlawcenter.com/state-minimum-wage-rates/.

17. Corkery, M. and S. Cowley. 2017. Household debt makes a comeback in the U.S. *New York Times*. May 17, https://www.nytimes.com/2017/05/17/business/dealbook/household-debt-united-states.html.

18. Jackson, T. 2009. *Prosperity Without Growth: Economics of a Finite Planet*. Oxon, UK: Earthscan.

19. McDonald, M.P. 2018. National general election VEP turnout rates 1789–present. *United States Election Project*. http://www.electproject.org/national-1789-present.

20. Editorial Board. Oct. 18, 2018. Voting should be easy. Why isn't it? *New York Times*. https://www.nytimes.com/2018/10/18/opinion/registration-vote-midterms.html?action=click&module=Opinion&pgtype=Homepage.

21. Pew Research Center. Nov. 23, 2015. Trust in government: 1958–2015. https://www.people-press.org/2015/11/23/1-trust-in-government-1958-2015/.

22. Putnam, R.D. 2000. *Bowling Alone: The Collapse and Revival of American Community.* New York, NY: Simon & Schuster.

23. Laozi (Lao-tzu, fl. 6th cn. B.C.E.). *Internet Encyclopedia of Philosophy.* https://www.iep.utm.edu/laozi/

24. Kessler, W. and A. Karlqvist. 2016. *Environmental Reality: Rethinking the Options.* Stockholm, Sweden: Royal Colloquium.

Glossary: Informal Definitions of Key Terms Used in This Book

Anthropocene New geologic interval, beginning in about 1950, that is dominated by human forces that shape Earth's future path.

Assisted migration Deliberate relocation of a species to a new location outside its historical range.

Biosphere All of Earth's nature—all of its ecosystems.

Bounded rationality Idea that individuals make decisions based on what they know or believe rather than a full understanding of the facts and issues.

Bridging organization Organization that shares information and coordinates activities of other groups that might not interact under normal circumstances but that can accomplish more and different things by working together than each could on its own.

Bycatch Non-target fish that are required by law to be thrown overboard.

Carbon dioxide (CO$_2$) Heat-absorbing gas, whose increase in the atmosphere is responsible for about two-thirds of recent climate warming.

Collaborative governance Collaboration of private citizens and other groups with government to seek consensus in making public decisions.

Collective action Joint effort made by a group to achieve a common goal.

Commons Lands, waters, and air that belong to and are used in common by society.

Cultural services Non-material benefits that society receives from ecosystems (for example, cultural identity, spiritual connections, aesthetics, recreation, and ecotourism opportunities).

Ecological solidarity (or social-ecological stewardship) Management approach that seeks to restore or enhance ecosystem health and human well-being.

Ecosystem All of the interacting components of nature at scales ranging from a termite's intestine to the Amazon basin.

Ecosystem degradation Deterioration of an ecosystem to a less desirable state as a result of failure to adapt or transform in response to direct and indirect human impacts.

Ecosystem health Metaphor that describes an ecosystem's condition, relative to its condition without negative human impacts.

Ecosystem services Benefits that society receives from ecosystems.

Ecosystem stewardship See Stewardship.

Empathy Understanding and sensitivity to the experiences of other people and of nature. It involves taking the perspective of others and feeling an emotional bond with them.

Environmental hazards Environmental stresses and events that harm people's health or well-being.

Externality A side effect or consequence of a commercial activity that affects other people, without this being reflected in the cost of the goods or services produced.

False dichotomy Complex sets of issues that are described simplistically as a choice between goals that are emblematic of different worldviews—for example, jobs versus the environment or climate action versus the economy.

Flourishing Favorable social and ecological outcomes in which society is empowered to define, shape, invent, and achieve its own goals.

Food insecurity Too little affordable nutritious food.

Food security Security that people's food needs are being met.

Formal education Classroom education.

Free riders People or institutions that benefit from exploiting a public good or service without paying for it.

Governance Processes of governing.

Greenhouse gases Atmospheric gases like carbon dioxide (CO_2) and methane (CH_4) that trap heat and warm the climate.

Greenwashing An unsubstantiated claim of the environmental benefits of a practice or product.

Grit Perseverance and passion to address long-term goals.

Gross domestic product (GDP) Monetary value of all the goods that a country produces. It is roughly equivalent to the total income earned or the value of the goods consumed.

Gross national happiness (GNH) An index of the happiness of a country's citizens, based on fair socioeconomic development, a vibrant culture, environmental protection, and good governance.

Hardpan Impermeable soil layer.

Homo economicus Fictional species of rational people who make conscious choices that are consistent with their self-interests and reflect their social and material preferences.

Homo environmentalis Fictional species that makes conscious choices based on desired social and environmental outcomes.

Homo materialis Fictional species that makes conscious choices based on material needs and desires.

Honorable consumption Consumption that is consistent with people's stewardship values.

Hubris Overconfidence in human capacity to plan for the future.

Human beings (or people in common terms) Members of the species *Homo sapiens*.

Humanity Collective human species, whose interactions with nature reflect a diversity of cultures and values.

Identity Qualities and beliefs that make a person or group different from others.

Impoverished Social and ecological outcomes that fail to meet sustainability goals.

Inalienable right A right that cannot be taken away.

Incompletely theorized agreement Agreement among people with clashing worldviews about a value that they share but which is only partially tied to their worldviews.

Induced traffic demand The increase in traffic that results from providing more or better roads.

Informal education Education that occurs outside the classroom.

Institution Organization or set of rules that guides people's behavior.

Isotopes Forms of an element such as carbon that differ in their atomic weight but behave in chemically similar ways.

Market failures Situations where unregulated markets reduce societal well-being.

Market imperfections Choices that do not contribute to personal satisfaction.

Median household income Half of the households earn more than this income, and half earn less.

Nature Living skin of planet Earth. It includes all of Earth's organisms and the water, air, soil, and rocks with which they interact.

Nitrogen-fixing plants Plant species whose associated bacteria convert atmospheric nitrogen to forms that plants can absorb.

Nudge Intervention such as an information campaign or hazard warning that is intended to alter people's behavior in a predictable way without explicit mandates or economic incentives.

Path dependence Dependence of future outcomes on previous events.

Permafrost Soil and rocks that remain frozen for at least 2 years.

Planned obsolescence Design of a product so that it will last only a limited time.

Polluter-pays principle Principle that a company should pay for the damage caused to society and the environment by the pollution that it releases.

Provisioning services Products that are directly harvested from ecosystems (for example, food, water, and fiber).

Regulating services Capacity of ecosystems to buffer disturbance and shape interactions among ecosystems (for example, regulating the climate, cleaning our drinking water, reducing disease risk, dampening storm waves and flooding).

Resilience Capacity of a system to sustain or develop its fundamental characteristics despite shocks and perturbations. This occurs by sustaining a diversity of response options.

Self-efficacy Belief that one can change one's life circumstances.

Sense of place People's attachment to and dependence on a place, as well as the meanings, values, and feelings that they associate with a place.

Social norms Attitudes, rules of behavior, and choices that we think characterize "our group."

Society Group of people living together in a more or less ordered community at scales ranging from a neighborhood to an assembly of nations.

Stakeholder People and other species that are affected by an organization's decisions.

Stewardship Human actions that shape Earth's future to restore or enhance ecosystem health and human well-being.

Stromatolite A calcareous mound built up of layers of lime-secreting blue-green algae and sediments. These ancient ecosystems produced enough oxygen to change the metabolism of planet Earth.

Sustainability People's relationship with nature in which they draw no more from nature than it can supply over the long run. Linked social-ecological systems are sustainable if they meet current human needs without compromising the ability of future generations to meet their needs.

Synergy Choices that provide mutual benefits.

Thriving Social and ecological outcomes that meet sustainability goals.

Trade-offs Choices that have both advantages and disadvantages.

Well-being (or quality of life) Health, happiness, and self-satisfaction of an individual or group of people.

Wicked problem A problem so complex that partial solutions create unexpected new problems.

Worldviews Ways that people think the world works.

Sources for Table and Figures

TABLE 3.1
Millennium Ecosystem Assessment. 2005. *Ecosystems and Human Well-being: Synthesis*. Washington, DC: Island Press. Table modified by Stuart Chapin and used with permission from the World Resources Institute.

FIGURE P.1
Top photo by Ray Atkeson (image 3155A of the Ray Atkeson Image Archive), courtesy of Rick Schafer Photography. Bottom photo (image ID-3749) taken by Jim Wark of Airphoto in 2003.

FIGURE 1.1
Chapin, F.S., III, P.A. Matson, and P.M. Vitousek. 2011. *Principles of Terrestrial Ecosystem Ecology*. 2nd edition. New York, NY: Springer. Figure modified by Stuart and Melissa Chapin and used with permission from Springer Nature.

FIGURE 2.1
Top photo by NASA, courtesy of Wikimedia Commons (https://en.wikipedia.org/wiki/Aral_Sea). Bottom photo courtesy of Shoista Agzamova of MICE. Uzbekistan.uz.

FIGURE 2.2
Left photo of Hans Jenny by Stuart Chapin in 1970. Right diagram modified by Stuart and Melissa Chapin from figure 2 in the unpublished report *The Pygmy Forest Ecological Staircase: A Description and Interpretation* prepared by Hans Jenny in 1973. [Pygmy Forest Jenny 1973—Mendocino Coast Recreation and]

FIGURE 2.3
Top photo courtesy of the National Oceanic and Atmospheric Administration Central Library, George E. Marsh Album (image ID theb1365). Bottom photo taken by Dorothea Lange near Lordsburg, New Mexico, in 1937 (image LC-DIG-fsa-8b38635 in the Farm Security Administration Collection at the US Library of Congress).

FIGURE 2.4
Top photo taken by Mark Dimmitt, courtesy of the Arizona-Sonora Desert Museum. Bottom photo taken by Brad Sutton, courtesy of the Joshua Tree National Park.

FIGURE 2.5
Photo courtesy of Eric Tourneret.

FIGURE 3.1
Chapin, F.S., III, G.P. Kofinas, and C. Folke (editors). 2009. *Principles of Ecosystem Stewardship: Resilience-Based Natural Resource Management in a Changing World.* New York, NY: Springer. Figure modified by Stuart and Melissa Chapin and used with permission from Springer Nature.

FIGURE 4.1
Top image courtesy of Camille St. Onge of the Washington State Department of Ecology. Bottom image courtesy of the National Oceanic and Atmospheric Administration (https://www.esrl.noaa.gov/gmd/ccgg/trends/).

FIGURE 4.2
Joel Pett editorial cartoon used with the permission of Joel Pett and the Cartoonist Group. All rights reserved.

FIGURE 4.3
Wuebbles, D.J. et al. (editors). 2017. *Climate Science Special Report: Fourth National Climate Assessment,* Vol. I. Washington, DC: US Global Change Research Program. https://science2017.globalchange.gov. Diagram modified by Melissa Chapin; data from Mann, M.E. et al. 2008. Proxy-based reconstructions of hemispheric and global surface temperature over the past two millennia. *Proceedings of the National Academy of Sciences* 105(36):13252–13257. https://www.pnas.org/content/105/36/13252.

FIGURE 4.4
Photos by Stuart Chapin (top) and by Diana Haecker of the Nome Nugget (bottom).

FIGURE 5.1
Original diagram by Stuart and Melissa Chapin. Information for constructing the diagram came from Coulthard, S., J.A. McGregor, and C. White. 2018.

Multiple dimensions of wellbeing in practice. Pp. 243–256. *In* Schreckenberg, K., G. Mace, and M. Poudyal (editors). *Ecosystem Services and Poverty Alleviation: Trade-offs and Governance.* London, UK: Routledge; Selva, J. 2019. Abraham Maslow, his theory and contribution to psychology. *Positive Psychology Program.* https://positivepsychologyprogram.com/abraham-maslow

FIGURE 6.1
Photos from Stuart Chapin's photo album.

FIGURE 6.2
Top photo courtesy of Greening of Detroit. Bottom photo courtesy of Michael Bryson of Roosevelt University.

FIGURE 6.3
Chapin, F.S., III, G.P. Kofinas, and C. Folke (editors). 2009. *Principles of Ecosystem Stewardship: Resilience-Based Natural Resource Management in a Changing World.* New York, NY: Springer; panels from Steffen, W. et al. (editors). 2004. *Global Change and the Earth System: A Planet Under Pressure.* New York, NY: Springer-Verlag. Figure modified by Stuart and Melissa Chapin and used with permission from Springer Nature.

FIGURE 6.4
Ecological footprint. *Global Footprint Network.* https://www.footprintnetwork. org/our-work/ecological-footprint/. Map courtesy of Jerrad Pierce. Permission provided by Creative Commons and by Jerrad Pierce.

FIGURE 6.5
Wynes, S. and K.A. Nicholas. 2017. The climate mitigation gap: Education and government recommendations miss the most effective individual actions. *Environmental Research Letters* 12:074024. https://iopscience.iop. org/article/10.1088/1748-9326/aa7541. Original diagram by Stuart and Melissa Chapin.

FIGURE 8.1
Ansell, C. and A. Gash. 2007. Collaborative governance in theory and practice. *Journal of Public Administration Research and Theory* 18:543–571. https:// academic.oup.com/jpart/article-abstract/18/4/543/1090370. Diagram modified substantially by Stuart and Melissa Chapin. Used with permission from Oxford University Press.

FIGURE 8.2
Photo courtesy of the Cleveland Press Collection, Michael Schwartz Library, Cleveland State University.

FIGURE 9.1
Ehle, J.M. 1965. *The Free Men*. New York, NY: Harper & Row. Photo courtesy of Jim Wallace (photographer and copyright holder). It is archived at the Wilson Library at the University of North Carolina.

FIGURE 10.1
Cartoon by Walt Kelly in 1971 for the first Earth Day poster.

Index

Tables, figures and boxes are indicated by *t*, *f* and *b* following the page number.